Ecological Diversity

Ecological Diversity

E. C. PIELOU
Dalhousie University
Halifax, Nova Scotia

A WILEY-INTERSCIENCE PUBLICATION
John Wiley & Sons New York • London • Sydney • Toronto

Library of Congress Cataloging in Publication Data

Pielou, E C 1924–
Ecological diversity.

"A Wiley–Interscience publication."
Bibliography: p.
Includes indexes.
1. Biotic communities—Mathematical models.
2. Ecology—Mathematical models. I. Title.

QH541.15.M3P53 574.5′24′0184 75–9663
ISBN 0–471–68925–4

Printed in the United States of America

10 9 8 7 6 5 4 3 2 1

Preface

The richness and variety—in a word, the diversity—of natural ecological communities have never been more highly valued than they are now, as they become increasingly threatened by the environmental crisis. Students of what has come to be known as "ecological diversity" realize that their work now has practical importance (indeed, urgency) in addition to the academic interest it has always had. It therefore seemed to me that a short state-of-the-art book would be useful to the many research workers and graduate students who are concerned with, or about to start work upon, this topic.

The central problem that students of ecological diversity are engaged on is easily stated. It is this: given a stable many-species community, how have its constituent species come to share the same habitat and how do they maintain themselves and interact with one another? Hence what determines the number of species that can live together, and their relative proportions? This is an enormous problem, compounded of numerous subproblems, and obviously it can be attacked from many different sides and by a variety of methods. Where the boundaries of the subject should be drawn is a matter of personal choice; the chapter headings will show how I have chosen. I have tried to cover the material thus selected concisely; my object is to guide those entering the field right to the center of it as quickly as possible. Then they can go to work without delay on a score of problems that seem on the brink of solution.

The level of mathematical difficulty differs in different parts of the book, just as it differs in different parts of the subject. This is, of course, unavoidable. A unified body of ecological theory is often arrived at by using an array of unrelated mathematical techniques and it is the motive for the work—an advance in ecology—that gives it unity, not the techniques employed.* There is no need, however, to read the chapters in order. Their natural grouping, by chapters, is as follows: (1), (2, 3, 4), (5, 6), (7, 8). But except for Chapter 4, which assumes that the reader has read Chapters 2 and 3, all the chapters are fairly self-contained and can be read in isolation.

The book was written while I held a Killam Research Professorship in the Biology Department of Dalhousie University.

E. C. PIELOU

Halifax, Nova Scotia
January 1975

* For an elementary account of some of those parts of the subject that *can* be covered at an elementary level, see the author's *Population and Community Ecology: Principles and Methods*, Gordon and Breach, 1974.

Contents

Ecological Diversity

Introduction

Any assemblage of plants and animals living together in one place and, to a greater or lesser degree, interacting with one another—in a word, an ecological community—automatically conjures up the following questions, among others, in the mind of an ecologist:

- Why are there not more (or fewer) species present, and will the number remain the same?
- Why are some of the species abundant and some rare, and are their relative proportions likely to remain fairly constant?
- Are most or all of the community's species fully adapted to the habitat they occupy and to one another?
- Which of the species are autochthonous (evolved locally) and which allochthonous (evolved elsewhere) and which (if any) will soon become locally or globally extinct?
- Do the species differ much in the amplitudes of their tolerance ranges for various environmental variables?
- How many trophic levels are present, and are there a small number of intricately anastomosing food webs or a large number of simple unbranched food chains?
- How many (and which) species could be removed without producing any marked effect, favorable or unfavorable, on the welfare of the others?

The foregoing is only a representative sample of the questions that would suggest themselves to a community ecologist; it would be easy, but pointless, to extend the list. However, even if we knew the answers to all these questions for a given community, there would still remain a more inclusive and basic problem: namely, are there any discoverable general laws governing the composition and structure of many-species communities, and if so what are they?

The last question is the preoccupation of students of community diversity. It is, obviously, one of the fundamental problems of theoretical ecology, the subject which bears to the other life sciences the same relation that cosmology bears to the other physical sciences. Obviously, also, it is far too monolithic a question to tackle head-on. Devising research projects that are both feasible, and relevant to the central problem, is one of the chief tasks of those who wish to contribute to diversity studies, and the object of this book is to survey the field and bring together in a single narrative a condensed account of the several lines of approach that appear to promise interesting advances.

In its recent rapid growth, theoretical ecology has split into two parts: mathematical ecology, and statistical ecology. They used to be treated as one, but now both parts are so large that one person can labor for a working lifetime in one of the parts without finding the field restricted. In studying diversity, both groups of workers have the same motive, that of uncovering general laws of community structure, but they use quite different methods.

Mathematical ecologists devise dynamic models, sets of differential-difference equations for instance, which are intended to simulate as realistically as possible the rates of the various interdependent population processes going on in a community. To do this they start from sets of apparently reasonable postulates and hope, by deduction, to arrive at conclusions that match events in the real world and presumably explain them.

Statistical ecologists as a rule have less faith in conceptual models and the long chains of argument arising from them. They prefer to take the world as they find it. However the statistical method in ecology does not (as some suppose) consist merely in amassing quantities of data and subjecting it to routine tests. It consists in looking at nature with a statistically informed eye; judging what questions can be asked with some assurance of arriving at unambiguous answers; deciding what data are needed and how they should be collected and analyzed to yield the answers; and then acting on these decisions.

Both approaches have drawbacks. The mathematicians run the risk of constructing interesting models divorced from reality; and the statisticians of providing clear answers to ecologically uninteresting questions. Perceiving what the reality is, and what questions must be answered if we are to understand it are the tasks of ecologists generally; no amount of mathematical or statistical expertise is any use if it is misapplied, and only an ecologist with considerable field experience can recognize good questions and good answers.

At the moment (1974) the literature on diversity is in spate. There is an information explosion and a speculation explosion and whether, between them, they will bring a breakthrough to any satisfyingly general theory remains to be seen. Speculation alone certainly will not; we have been told often enough how a community *should* behave given the correctness of this or that hypothesis. Hypotheses abound. The difficulty is to test them.

The current worldwide climatic deterioration may provide the tests we need; it may plunge us all into a vast, though undesired, ecological field experiment. If we observe the concomitant changes in the biosphere we may end up sadder and wiser in a very literal sense.

Chapter 1

Indices of Diversity and Evenness

1.1 Introduction

All ecologists are familiar with the variability of natural communities. Some communities (e.g., the mollusks of a rocky shore in a cold climate) may consist of only a few species whereas others (e.g., the trees in a tract of tropical rain forest) may have hundreds of species. In this respect, therefore, as well as in many others, communities exhibit a property in which they vary enormously and it is convenient to name this property *diversity*. It bears to qualitative observations the relationship that *variance* bears to quantitative measurements. In the same way that statistical variance provides a measure of, for instance, the variability in height of the trees in a forest, a diversity index measures the variability of the trees' species identity.

The purpose of measuring a community's diversity is usually to judge its relationship either to other community properties such as productivity and stability, or to the environmental conditions that the community is exposed to. We defer to later chapters a consideration of the conclusions that "diversity studies" have yielded. In this chapter the measurement of diversity is discussed.

Before diversity can be measured it is obvious that we must define precisely the collection of organisms that comprises the community concerned. Communities vary tremendously: one might define as a community the sea birds inhabiting a rocky island in the breeding season; the arthropods in a rotting log; the ungulates in an area of thousands of square kilometers of African grassland in the rainy season; the catch of plankton in a single tow of a net; the plants in a tract of bog; or the aquatic invertebrates caught in a Surber sampler. Exact descriptions of these (or any) communities must specify (1) the spatial boundaries of the area or volume containing the community and the way in which sampling was done; (2) the time limits between which observations were made; and (3) the taxocene* treated as constituting the community. For two of these examples (the catch in a plankton net, and the catch in a Surber sampler) the word "community" may seem unwarranted since the individual organisms comprising them may not, as a group, have constituted ecological entities before they were caught. Nevertheless, one may wish to know the diversities of such assemblages and they can be classed as communities for convenience.

Unless diversity is to be defined simply as the number of species present (as it sometimes is), it is also necessary to decide upon a method for measuring the amount of each species in a community. For many animals, and for plants that do not reproduce vegetatively, the number of individuals is the obvious measure. But if the individuals vary greatly in size, as they may when the taxocene concerned contains widely dissimilar species or individuals of one species of widely varying ages, the biomass of a species is a more reasonable measure of its quantity. Biomass (or alternatively, dry weight) is also a suitable measure if the species, instead of occurring as countable individuals, form extensive patches as do vegetatively reproducing plants and many colonial coelenterates. For organisms such as cushion plants, lichens, and crustose bryozoa, "cover" is a good measure of quantity.

In what follows it is assumed for convenience that species quantities are

* The members of some chosen taxon occurring together. The taxon may be of any (supraspecific) rank. The term is attributed (by Hutchinson, 1967) to Chodorowski.

expressed as numbers of individuals. The arguments would be unaltered if some other measure of quantity were used instead.

1.2 An Index of Diversity

As stated above, one measure of a community's diversity is simply the number of species it contains, say s. This, however, is an unweighted measure analogous to the range (in the statistical sense) of a quantitative variate. To arrive at a weighted measure, that is, one that takes account of the relative quantities of the s species, we require some appropriate function of the proportions p_i $(i=1, \ldots, s)$ where p_i is the proportion of the community belonging to the ith species. We are free to choose what this function, say $H'(p_1, \ldots, p_s)$, shall be.

Three desirable properties for H' are the following.

1. For given s, H' should have its greatest value when $p_i = 1/s$ for all i. Such a community will be called completely even.
2. Given two completely even communities, one with s species and the other with $s+1$, the latter should have the greater H'.
3. This property is needed when the individuals in a community are classifiable in more than one way. It is this:

Suppose the community members are subjected to two separate classifications (not necessarily independent), namely an A-classification with a classes and a B-classification with b classes. Let p_i $(i=1, \ldots, a)$ be the proportion of community members in the ith class of the A-classification; let q_{ij} $(i=1, \ldots, a;\ j=1, \ldots, b)$ be the proportion of *these* members that belong to the jth class of the B-classification. And put $p_i q_{ij} = \pi_{ij}$ so that π_{ij} is the proportion of the whole community that belongs to the ith A-class and the jth B-class.

Also, put $H'(AB)$ for the diversity of the doubly classified community; $H'(A)$ for the diversity under the A-classification only; and $H'_i(B)$ for the diversity under the B-classification of that part of the community belonging to the ith A-class.

Let $H'_A(B) = \sum p_i H'_i(B)$ be the mean of the $H'_i(B)$ over all A-classes.

We then require that

$$H'(AB) = H'(A) + H'_A(B). \tag{1.1}$$

It should be noticed that if the A- and B-classifications were independent so that $q_{ij} = q_j$ for all i and $\pi_{ij} = p_i q_j$, then

$$H'_i(B) = H'(B) \qquad \text{for all} \quad i;$$
$$H'_A(B) = H'(B) \sum p_i = H'(B);$$

and

$$H'(AB) = H'(A) + H'(B).$$

Of the three properties listed above, 2 is obviously desirable in an index of diversity. Property 1 ensures that an "even" community shall have a greater index of diversity than one in which, though the number of species is the same, the community is dominated by one or a few of them in which case its diversity in the intuitive sense would be less. The usefulness of property 3 will be discussed in Section 1.7.

Now it can be shown (Khinchin, 1957; Pielou, 1969) that the only function of the p_i values having these three properties is

$$H'(p_1, \ldots, p_s) = -C \sum p_i \log p_i$$

where C is a positive constant. Putting $C = 1$, we may therefore take, as an index of diversity,

$$H' = -\sum p_i \log p_i \,. \tag{1.2}$$

This index, which is now widely used by ecologists, has come to be known as the Shannon (or Shannon–Wiener) index. It was originally proposed by Shannon (Shannon and Weaver, 1949) as a measure of the information content of a code.

1.3 The Simpson Index

The Shannon index defined in (1.2) is a special case of a more general class of functions used in the mathematical theory of information. Given a code of s kinds of symbols, of which a proportion p_i are of the ith kind, the function

$$H_\alpha = \frac{\log \sum p_i^\alpha}{1-\alpha} \tag{1.3}$$

is known as the entropy of order α of the code (equivalently, of the set $\{p_i\}$); (*see* Renyi, 1961).

It will be seen (using l'Hospital's rule) that

$$H_1 = \lim_{\alpha \to 1} H_\alpha = -\sum p_i \log p_i = H'.$$

Thus H' is identical with the entropy of order 1 of the set $\{p_i\}$.

Next consider H_2. Putting $\alpha = 2$ in (1.3) gives

$$H_2 = -\log \sum p_i^2. \tag{1.4}$$

The function $\sum p_i^2 = \lambda$, say, has been used as an index of the "concentration" or "dominance" of a many-species community. The measure was first proposed by Simpson (1949) who described λ as a measure of concentration. The reason for doing so is that λ is the probability that any two individuals picked independently and at random from the community will belong to the same species. It thus measures a property that is the opposite of diversity: obviously, the greater the chance that two randomly picked individuals are of the same species, the less a community's diversity in the intuitive sense.

The function λ is also a measure of the "expected commonness" of an event (Weaver, 1948). In the present context, consider the event: a randomly picked individual is found to belong to species i. Define as a measure of the *commonness* of this event the value p_i, which is its probability of happening. Then $\lambda = \sum p_i^2$ is evidently the expected commonness of a randomly chosen individual. If a single species, say species 1, "dominates" a community (so that the community's diversity is low) then p_1, which is both the species' commonness and its probability of being chosen, is near unity and λ is high. Conversely, if there are numerous species all fairly evenly represented, all are "uncommon" and λ is low.

This suggests the use of some function of λ that increases with decreasing λ as an index of diversity. One possibility (Pielou, 1969) is to use $1-\lambda$. A better function* is $-\log \lambda = H_2$ since this emphasizes its connexion with $H_1 = H'$. Both H' and $-\log \lambda$ are "entropies," of order 1 and 2 respectively. (This fact, though mathematically interesting, is irrelevant to ecology of course; it cannot be too strongly emphasized that fancied links between the information-theoretic concept of "information" and the diversity of an ecological community are merely fancies and nothing more.)

At present both H' and functions of λ are used by ecologists as diversity indices. However, only the former possesses property 3 given in (1.1) and this makes it a more useful index (see Sections 1.7 and 8.3). It will be used exclusively in what follows.

The units in which diversity is measured depend on the base of the logs (Pielou, 1969): with logs to base 2 the unit is a *binary digit* or *bit*, with logs to base 10 the unit is a *decit*, and with natural logs a *nat*.

1.4 The Diversity of Fully Censused Communities

Communities whose diversities are of interest are occasionally small enough to be censused in their entirety. When this is so, diversity can be

* I am indebted to Professor C. D. Kemp for pointing this out.

determined without error from complete knowledge, instead of being estimated (with error) from the incomplete knowledge yielded by a sample.

The following are examples of communities that might be fully censused: the breeding birds in a small tract of forest; the shrubs and trees that have established themselves in an abandoned quarry; the catch in an insect light trap. Communities as small as these are of little interest in isolation or if observed only once; but data from a small community observed at a sequence of times are often required in studies of succession and trends.

In what follows we shall call a fully censused community a *collection*. The Shannon index H' discussed in the preceding section is not appropriate for measuring a collection's diversity; it should be used only for "indefinitely large" communities, that is, communities that can be treated as infinite in the sense that removing samples from them causes no perceptible change in them. For a collection, the appropriate measure of diversity is Brillouin's index (Brillouin, 1962; Margalef, 1958) defined as

$$H = \frac{1}{N} \log \frac{N!}{\Pi N_i!} \tag{1.5}$$

where N is the number of individuals (or other units of quantity) in the whole collection and N_i is the number in the ith species for $i = 1, \ldots, s$.

It should be noticed that H (without prime) and H' (with prime) will be used for the Brillouin index and the Shannon index respectively. H, as well as H', has the three desirable properties of a diversity index listed in Section 1.2.

The connexion between H and H' is this. In the mathematical theory of information H' measures the information content per symbol of a code or language; from such a code indefinitely many messages may be constructed (or "samples removed") without depleting the code itself. The Brillouin index H measures the information content per symbol of a particular message (Goldman, 1953). By analogy we may treat a large community (that has to be sampled rather than censused) as a "code" and a collection as a "message."

Analogies with information theory are instructive but do not, of course, provide a compelling reason for using H' and H in the way just outlined. Resemblances between codes and messages on the one hand, and communities and collections on the other are only superficial. However, there are two far stronger reasons for using H' for communities and H for collections. These are as follows.

1. Use of different indices in the different circumstances emphasizes the contrast between them. A value of H is determined from a complete census

and hence is free of statistical error whereas a value of H' is estimated (as opposed to "determined") from the incomplete knowledge yielded by a sample and thus has a sampling error; estimates of H' should always be accompanied by estimates of their standard errors, as discussed in the next section.

2. Given a finite s-species collection of N individuals with N_i in species i (whose diversity is defined as H), and an infinite community with proportion $N_i/N = p_i$ of its individuals in species i (whose diversity is defined as H'), it is always true that $H' > H$. This follows since

$$H' - H = \frac{-1}{N} \log \left[\frac{N!}{N_1! \cdots N_s!} \left(\frac{N_1}{N} \right)^{N_1} \cdots \left(\frac{N_s}{N} \right)^{N_s} \right].$$

The term in brackets is a multinomial probability and hence less than unity. Therefore $H' - H > 0$. It follows that if, from an indefinitely large community, we take two samples, one small and one large, and treat both as collections, the small collection would be expected to have a lower value of H than the large collection. This result accords with what we intuitively require of a diversity index as will become clear below (Chapter 3).

It should finally be noticed that

$$\lim_{\min(N_i) \to \infty} H = H'.$$

This may be demonstrated using the relation

$$\lim_{N \to \infty} \log N! = N \log N - cN \tag{1.6}$$

where $c = \log e$. If natural logs are used in (1.6), $c = 1$.
[*Note:* The approximation to $\log N!$ given in (1.6) is not good for practical calculations. Unless N is very large indeed the approximation is not at all close.]

Then

$$\begin{aligned} H &\to \frac{1}{N} (\log N! - \sum \log N_i!) \\ &= \frac{1}{N} [(N \log N - cN) - (\sum N_i \log N_i - c \sum N_i)] \\ &= -\sum \frac{N_i}{N} \log \frac{N_i}{N} = H'. \end{aligned}$$

1.5 Estimating the Diversity of a Large Community

For large communities, diversity must be estimated from a sample. Suppose one gathers a sample of n sampling units (s.u.'s) and determines the amount of each species in each s.u. and hence in the whole sample.

Let N_i be the amount of the ith species in the whole sample and let $\sum N_i = N$.

Although $\hat{p}_i = N_i/N$ is the maximum likelihood estimator of the proportion of species i in the whole community, it is *not* desirable to estimate H' by putting

$$\hat{H}' = -\sum \hat{p}_i \log \hat{p}_i.$$

This is so for two reasons: (1) The estimator $\hat{H}'$ is biased (Basharin, 1959). The bias may be allowed for, but only if one knows the total number of species in the whole community, say s^*. Often this number is not known and one can say only that s, the number of species in the sample, gives a lower bound for it. (2) Use of $\hat{H}'$ entails the tacit assumption that the s.u.'s examined constitute a random sample from the community whose diversity is being estimated. But this is very difficult (probably impossible) to ensure. It is true that the s.u.'s can be randomly located throughout the area (or volume) occupied by the community but this provides a random sample of space, not of the organisms occupying it, whose pattern is almost certainly patchy. The following is a way of dealing with the difficulty (Pielou, 1966).

List the s.u.'s and their contents in random order and rank this random ordering from 1 to n. Let the amount of species i in the xth s.u. be N_{xi}. Put

$$M_{ki} = \sum_{x=1}^{k} N_{xi} \qquad \text{and} \qquad M_k = \sum_{i=1}^{s} M_{ki}.$$

Also, put

$$\mathscr{H}_k = \frac{1}{M_k} \log \frac{M_k!}{\prod M_{ki}!}.$$

Thus M_{ki} is the quantity of species i in the pooled contents of the first k of the randomly ordered s.u.'s; M_k is the quantity of all species in these pooled s.u.'s; and $\mathscr{H}_k$ is their diversity (using Brillouin's index).

Now plot $\mathscr{H}_k$ versus k; (Figure 1.1 shows examples). It will be found that $\mathscr{H}_k$ tends to increase with k (not necessarily monotonically) and, provided n is large enough, will level off when (intuitively speaking) enough s.u.'s have been added to the pool to yield a "representation" of the whole community.

Suppose the curve is effectively horizontal (i.e., shows no upward trend) as soon as t s.u.'s have been pooled. Put

$$h_k = \frac{M_k \mathscr{H}_k - M_{k-1} \mathscr{H}_{k-1}}{M_k - M_{k-1}}.$$

Then each h_k for all $k > t$ can be treated as a separate estimate of the community diversity H'. To see this, suppose we had censused the whole

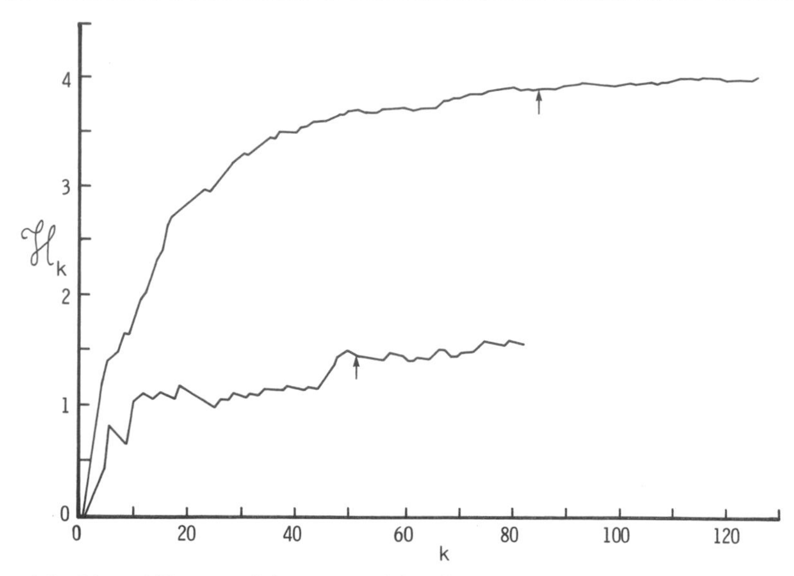

Figure 1.1 Plots of $\mathcal{H}_k$ versus k for communities of amphibians and reptiles in moist tropical forest in Ecuador (upper curve) and dry evergreen forest in Thailand (lower curve). The arrows show the positions of t. (Redrawn from Heyer and Berven, 1973.)

community and had denoted its diversity by

$$H_T = \frac{1}{N} \log \frac{N!}{\Pi N_i!}.$$

Now put $B_T = NH_T$ for the community's total diversity (as opposed to its diversity per individual). Next, imagine that a single individual chosen at random from the community is removed. Since, with probability N_i/N, it belongs to species i, the expected total diversity of what remains after its removal is

$$E(B_{T-1}) = \sum \frac{N_i}{N} \log \frac{(N-1)!}{N_1! \cdots (N_i - 1)! \cdots N_s!}.$$

Therefore,

$$B_T - E(B_{T-1}) = \sum \frac{N_i}{N} \left[\log \frac{N!}{N_1! \cdots N_i! \cdots N_s!} \cdot \frac{N_1! \cdots (N_i - 1)! \cdots N_s!}{(N-1)!} \right]$$

$$= -\sum \frac{N_i}{N} \log \frac{N_i}{N}.$$

Thus the expected result of removing a single individual is to reduce the total diversity by an amount H'. Now apply the same arguments to the pooled contents of $(k+1)$ s.u.'s and assume k sufficiently large for the pooled s.u.'s to be treated as "representative" of the entire community. If the $(k+1)$th s.u. contains only one individual, the expected result of removing this s.u. will likewise be to reduce the pool's total diversity by the quantity H'. That is, we may put

$$E\left[\frac{M_{k+1}\mathcal{H}_{k+1}-M_k\mathcal{H}_k}{M_{k+1}-M_k}\right]=H' \qquad \text{when} \qquad M_k=M_{k+1}-1.$$

This result is exact. For sufficiently large k the relation will still be approximately true even when the $(k+1)$th s.u. contains more than one individual and its contents are not a random sample from the whole community. Therefore, putting

$$\frac{M_{k+1}\mathcal{H}_{k+1}-M_k\mathcal{H}_k}{M_{k+1}-M_k}=h_{k+1},$$

we may assume $E(h_k)\simeq H'$ for all $k>t$ provided t is large enough. The required value of t must first be judged subjectively from inspection of the $\mathcal{H}_k$ versus k curve; one may then judge whether a sufficiently large t has been picked by testing whether successive values of h_k for $k=t, t+1, \ldots, n$ exhibit serial correlation. If they do not, it may safely be assumed that a pool of t or more s.u.'s is large enough for successive values of h_k to be regarded as independent. Then H' may be estimated by

$$\tilde{H}'=\bar{h}=\frac{1}{n-t}\sum_{k=t+1}^{n} h_k,$$

and

$$\tilde{\text{var}}\,(\tilde{H}')=\tilde{\text{var}}\,(\bar{h})=\frac{1}{n(n-1)}\left(\sum h_k^2-n\bar{h}^2\right).$$

Recall that this procedure for estimating H' begins with the random ordering of the s.u.'s. Hence a different value of H' will be obtained if the procedure is repeated. Clearly, $n!$ different outcomes are possible. Lloyd, Inger, and King (1968) recommend that if several replicate estimates are calculated, each based on a different permutation of the s.u.'s, the median result be taken as "definitive."

1.6 Evenness and Equitability

The diversity of a community depends on two things: the number of species and the evenness with which the individuals are apportioned among them. To describe a community's diversity merely in terms of its diversity index is to confound these two factors; a community with a few, evenly represented species can have the same diversity index as one with many, unevenly represented species. It is obviously desirable to keep distinct these two ingredients of diversity, the number of species, s, and the evenness. A method of measuring evenness is therefore required and we shall here consider three.

The first of these is to use the ratio of the observed diversity index to the maximum value the diversity index could have in a community with the same number of species. Denoting by J and J' the evenness in a "collection" (fully censused) and a "community" (sampled) respectively, we may put $J = H/H_{max}$ and $J' = H'/H'_{max}$.

Calculating J presents no difficulties. H_{max} is the diversity index of a hypothetical collection, used as a standard, having both the same number of species and the same number of individuals as the observed collection. Therefore unless N is a multiple of s, the hypothetical standard collection will not have the same number of individuals in all the species. Suppose

$$N = s\left[\frac{N}{s}\right] + r$$

where $[N/s]$ is the integer part of N/s and r is the remainder. Then putting $[N/s] = X$ and $X + 1 = Y$, so that $N = (s - r)X + rY$, it is seen that

$$H_{max} = \frac{1}{N} \log \frac{N!}{(X!)^{s-r}(Y!)^{r}}.$$

Since J relates to a collection it is determinable without sampling variance and hence has no standard error.

The calculation of J' is also straightforward provided we know s^*, the total number of species in the sampled community. Assume for the moment that all the community's species are present in the sample so that $s^* = s$. Then complete evenness implies that

$$H'_{max} = -\sum \frac{1}{s} \log \frac{1}{s} = \log s,$$

and

$$J' = \frac{H'}{\log s}.$$

Logs to the same base are used, of course, for both the numerator and the denominator of this dimensionless ratio. Any arbitrary base, say a, may be used. It then follows that (since $\log_s x = \log_a x/\log_a s$), division of H' by $\log s$ has the effect of altering the units in which H' is measured. Thus if, instead of expressing the diversity index in arbitrary units (such as bits, decits, or nats), we let the choice of units be decided by s and use logs to base s in calculating H', the resultant H' is identically equal to J', the evenness.

The obvious estimator of J' is $\tilde{J}' = \tilde{H}'/\log s$ with $\tilde{H}'$ obtained by the method described in Section 1.4. Since s is assumed known, $\log s$ is free of sampling error and

$$\widetilde{\text{var}}\,(\tilde{J}') = \frac{\widetilde{\text{var}}\,(\tilde{H}')}{(\log s)^2}.$$

If s is *not* known, as is very often the case, estimation of J' entails estimation of s^*. In most cases this is impossible. It is true that fitting a theoretical species abundance distribution to the corresponding observed distribution (see Chapter 3) sometimes yields an estimate of s^* as a by-product, but such an estimate does not usually inspire confidence. The conclusion seems unavoidable that there is no way of estimating the evenness of a large, uncensused community that is suspected of containing an unknown number of "odds and ends"—species that might be represented by one or two individuals but equally well might not. Such "incidental" species are usually of no ecological importance and could be disregarded; but then an arbitrary decision would have to be made on where the dividing line should be drawn between important and unimportant species.

A modified method of measuring evenness has been suggested by Hurlbert (1971) and DeBenedictis (1973). Consider a censused collection, again: its evenness as measured by J is not independent of H since no allowance is made for the fact that for given N and s there is a minimum as well as a maximum possible value for H. Clearly, $H_{\min}$ occurs when all but one of the s species is represented by a single individual only.

Then

$$H_{\min} = \frac{1}{N} \log \frac{N!}{(N-s+1)!}. \tag{1.7}$$

As a measure of evenness that does not depend on s one may therefore use V defined as

$$V = \frac{H - H_{\min}}{H_{\max} - H_{\min}}.$$

An analogous statistic, which could be called V', can*not* be defined for uncensused communities, even when s^* is known; (when s^* is not known, of course, the impossibility of determining $H'_{\max}$ makes V' as well as J' intrinsically inestimable). This follows since no clear meaning can be attached to $H'_{\min}$ for communities that are "indefinitely large" but, like real communities, somewhat less than infinite. From (1.7) it is seen that

$$\lim_{N \to \infty} H_{\min} = 0$$

and hence that $V' = J'$ identically.

The third measure of evenness, known as "equitability," is due to Lloyd and Ghelardi (1964). Instead of taking complete evenness as the standard against which the evenness of an observed community is to be judged, they take as standard the evenness of the so-called "broken stick" distribution. (The derivation and properties of the distribution are discussed in Chapter 2.) This theoretical distribution has, as its only parameter, s the number of species. The Shannon diversity index of a community whose species abundance distribution is given by the broken stick law, $M'(s)$ say, is therefore uniquely determined by s. Writing $H'(s)$ for the diversity of an observed s-species community, one might use the ratio $H'(s)/M'(s)$ as a measure of its evenness. (This ratio need not be less than one though it usually is; natural communities sometimes have species abundance distributions of greater evenness than that predicted by the broken stick law, but they are unusual.)

A more intuitively meaningful measure of evenness is derived by determining the value of s' for which $M'(s')=H'(s)$ and then taking, as the "equitability" of the observed community, the ratio $\varepsilon=s'/s$. Thus an s'-species community whose species abundances had the broken stick distribution would have the same (Shannon) diversity index as the observed s-species community. Usually, but not always, it is found that $s'<s$. Lloyd and Ghelardi provide a table of $M'(s')$ for a wide range of values of s'; the table can be used to find the required value of s' for calculating ε for an observed community.

1.7 Hierarchical Diversity

The diversity indices thus far considered take no account of taxonomic differences above the species level. But consider any s-species community with given proportions of its members in the several species. One would obviously regard its diversity as greater if the species belonged to several genera than if they were all congeneric, and as greater still if these genera belonged to several families than if they were confamilial. A way of measuring diversity that allows for the hierarchical nature of taxonomic classification is therefore desirable. Using the diversity indices H' (for communities) and H (for collections), this is easily done. The method uses property 3 mentioned in Section 1.2.

Suppose, for simplicity, that two taxonomic levels only are considered, genus and species: (the argument can easily be generalized to take account of more levels). Let the community contain g genera of which the ith contains s_i species ($i=1, \ldots, g$). Let q_i be the proportion of the community's individuals (not species) that belong to the ith genus; and let p_{ij} be the proportion of the ith genus's members that belong to species j in this genus ($j=1, \ldots, s_i$). Thus $q_i p_{ij}$ is the proportion this species forms of the whole community.

Now put $H'(SG)$ for the species-diversity of the whole community; $H'(G)$ for the genus-diversity of the whole community; $H'_i(S)$ for the species-diversity in the ith genus; and $H'_G(S)=\sum_{i=1}^{g} q_i H'_i(S)$ for the mean within-genus species-diversity averaged over all genera. Then

$$\begin{aligned} H'(SG) &= -\sum_{i=1}^{g}\sum_{j=1}^{s_i} q_i p_{ij} \log q_i p_{ij} \\ &= -\sum_i q_i \log q_i\Big(\sum_j p_{ij}\Big) + \sum_i q_i\Big(-\sum_j p_{ij} \log p_{ij}\Big) \\ &= H'(G) + H'_G(S). \end{aligned} \tag{1.8}$$

For three taxonomic levels, say family, genus, and species, the analogous relationship (with obvious symbols) is

$$H'(SGF) = H'(F) + H'_F(G) + H'_{GF}(S). \tag{1.9}$$

It is easily seen that these relationships also hold for the Brillouin index. Thus (1.8) and (1.9) show how a community's species-diversity as measured by the Shannon index may be divided into hierarchical components. And the same equations with the primes omitted show how a collection's diversity, measured by the Brillouin index, may be similarly divided; (and see Section 8.3).

Examples of the measurement of hierarchical diversity in a natural community will be found in Lloyd, Inger, and King (1968) in their account of communities of reptiles and amphibians in rain forest in Borneo.

Chapter 2

Species Abundance Distributions I

2.1 Introduction

Whenever a collection is censused or a community sampled, the resultant data consist of a list of the species in the collection or sample together with a statement of the quantity of each species. These quantities may be expressed as numbers of individuals, biomass values, dry weights, or cover values (see Section 1.1), as appropriate. For convenience we shall continue to assume that the amount of each species is measured by counting its individuals, but this in no way affects the generality of the arguments.

From such data diversity indices may be determined or estimated as described in Chapter 1. But a community's diversity index is merely a single descriptive statistic, only one of the many needed to summarize its characteristics and, by itself, not very informative. The belief (or superstition) of some ecologists that a diversity index provides a basis (or talisman) for reaching a full understanding of community structure is wholly unfounded. This is not to say that diversity indices are useless, only that they should not be overvalued. The analogy between a diversity index and a variance (see Section 1.1) should be borne in mind. It is obvious that detailed study of the distribution of a quantitative variate entails more than merely calculating its variance. In the same way, detailed study of the structure of an ecological community entails more than calculating its diversity index.

An obvious topic of interest in the study of communities is the distribution of the number of individuals per species. A community is often found to contain several similar species of apparently similar requirements, but with the species differing widely in their relative abundances. Sometimes the differences can be explained in terms of habitat differences: the abundance of a species may be proportional to the relative amount of space—in terms of suitable habitat—available to it. But even when this is the case, it gives only a proximate reason for the relative abundances of a community's member species. The ultimate reason would consist of an explanation of why the species have evolved with such widely different tolerance ranges. Investigating models that might account for observed species abundance relations is thus of great theoretical interest.

Ecologists tabulate species abundances in two quite different ways, as ranked-abundance lists and as species-abundance distributions. The method chosen depends on the number of species in the collection (or sample) that a body of data describes. If the data concern only a few species it is customary to list the number of individuals in each, ranking these numbers from largest to smallest; that is, one compiles a *ranked-abundance list.*

However, if the collection is very large and contains numerous species, of which several have exactly one individual, several have exactly two, and so on, the most convenient way of summarizing the data is as an observed frequency distribution giving the frequency, say f_r, of species represented by r individuals for $r=1, 2, \ldots$. The result is a *species-abundance distribution.*

The kinds of models that have been devised to explain the relative abundances of the species in a community show a similar (though not exactly corresponding) dichotomy. One kind may be called "resource apportioning models"; such a model is constructed by postulating the way in which coexisting species subdivide among themselves some necessary resource which is assumed to be the limiting factor (the same for all of them) that sets a limit to each species' population size. Three of these models are described in this chapter; two of them predict a community's ranked-abundance list (as defined above) and the third predicts a community's species-abundance distribution.

Models of the other kind will be called "statistical models." They consist of assumptions about the probability distributions of such variates as the numbers of members of each of several species in a defined area, and their predictions are expressed as species-abundance distributions. Three examples are given in Chapter 3.

As will be seen, the same predictions can be yielded by two or more models, and these may belong to both the contrasted kinds. It will also be shown how one may derive a model's species-abundance distribution from its ranked-abundance list.

2.2 The Niche Preemption Model

Suppose the dominant species in an s-species community preempts as its share a fraction k of the limiting resource, the second strongest species preempts a fraction k of what remains, and so on. Suppose also that the abundance of each species is proportional to the resource fraction it has preempted. Then the ranked-abundance list (as proportions) is k, $k(1-k)$, $k(1-k)^2, \ldots, k(1-k)^{s-2}$, $(1-k)^{s-1}$.

Observe that the abundances of the first $s-1$ species form a geometric series; and that the sth species occupies the resource-fraction $(1-k)^{s-1}$ that remains after the other species have been accommodated. To ensure that this last term is also the smallest, we must have $k \geq 0.5$.

It should also be noticed that it is the ranked-abundance list, *not* the species-abundance distribution that, except for the last term, forms a geometric series. Therefore it is misleading to describe the model as the "geometric series model" as is sometimes done (Whittaker, 1972; May, 1975). Indeed although the model predicts that, in a community of total size N, the expected number in the ith species will be $E(y_i) = Nk(1-k)^{i-1}$, $E(y_i)$ should not be regarded as an expected frequency but as a measure of species i's predicted share of the shared resource, or equivalently, as the predicted "size" of species i. Each of these sizes occurs with frequency one.

The species-abundance distribution that corresponds to this ranked-abundance list has been obtained by May (1975) as follows. We assume that there are sufficiently many species in the community for the distribution to be treated as continuous although in fact the variate (species "size" or abundance) takes only as many different values as there are species. We require to find the probability density function (pdf), say $\psi(y)$, of this variate. Visualize a plot of its cumulative distribution function (cdf). Figure 2.1*a* shows an example with $s=5$ and $k=0.6$; (this very low value of s has been used for clarity of illustration). The step function, with $s=5$ steps of equal height adding to unity, is the true, expected cdf, and the smooth curve is the graph of the approximating continuous function, $\Psi(y)$.

Thus Pr{a randomly chosen species has size $\geq y_i$} $\simeq 1-\Psi(y_i)$. But this probability is obviously proportional to the expected number of species whose sizes are greater than or equal to that of the ith species, and this is, by definition, equal to i, the rank of the species when they are ranked from largest to smallest (equivalently, when they are ranked from right to left on the abscissa of the graph). Now since, according to the model, the proportional abundance of this species is

$$y_i = k(1-k)^{i-1} \tag{2.1}$$

$$i = \frac{\log y_i - \log k(1-k)}{\log(1-k)}, \tag{2.2}$$

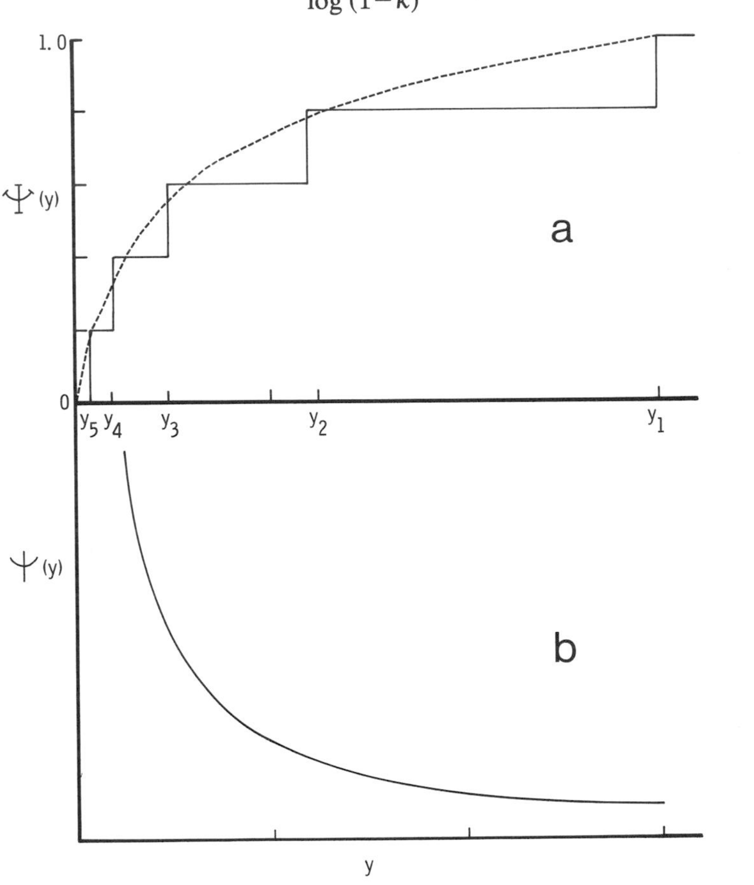

Figure 2.1 (a) The step function is the expected cdf of y, the relative abundances of the species, assuming a niche preemption model with $s = 5$, $k = 0.6$. An empirical cdf based on a sample containing all 5 species would have the locations of the steps shifted left or right because of sampling variation, but they would still be $s = 5$ in number and of equal height. The smooth curve (dashed line) is an approximation to the true cdf. (b) The pdf corresponding to the approximate (smooth) cdf in (a).

and hence

$$1-\Psi(y) \propto i = \frac{\log y}{\log(1-k)} + \text{const.}$$

The required pdf, $\psi(y)$, is then given approximately by

$$\psi(y) = \Psi'(y) \propto -\frac{d}{dy}\left(\frac{\log y}{\log(1-k)}\right)$$

so that

$$\psi(y) = \frac{C}{y} \qquad \text{for} \qquad (1-k)^{s-1} \le y \le k. \tag{2.3}$$

Here C is a constant of proportionality that must be such as to ensure that

$$\int_{(1-k)^{s-1}}^{k} \psi(y)\, dy = 1.$$

Figure 2.1*b* shows the graph of this pdf.

Evaluation of C and further investigation of $\psi(y)$ is probably not worthwhile. The niche preemption model is normally envisaged as a possible explanation of how a limiting resource might be divided up among a small number of species, typically 20 or fewer. Use of the continuous function $\psi(y) = C/y$ as an approximation to the true discrete distribution is only justifiable for large values of s but then the model underlying the distribution is not likely to be entertained in any case. It is interesting, however, to note the shape of $\psi(y)$. Subsequently (see Section 2.4) this pdf will be compared with those predicted by other models; it will be seen that the niche preemption model assumes that the members of the most abundant species (the dominant species) constitute a very large proportion of the whole community; in other words, that the degree of dominance of this species is very pronounced, more so, indeed, than in any other model.

An example of its application is given by Whittaker (1972) who assembled and compared observations on the "sizes" of the component species in a series of communities; (the size of a species was measured by its shoot productivity). The communities represented the stages in a vegetation succession from old-field communities to oak-pine forest (in Long Island, New York). Whittaker found that only in the earliest stage was the dominance of the most abundant species so pronounced or (which comes to the same thing) the evenness of the community so low, that the niche preemption model appeared to accord with the observations.

The problem of judging the fit of the model to actual observations is deferred to Chapter 4.

2.3 The Broken Stick Model

This model was first proposed by MacArthur (1957). Like the niche preemption model, this one also is a "resource apportioning model." The limiting resource that must be divided up is likened to a "stick," or line, of unit length that is broken into s disjunct segments by breaks at $s-1$ points located at random along its length. The lengths of the segments represent the "sizes" (in the sense used in the preceding section) of the species. According to the model, the expected size of the ith species, y_i, is then

$$E(y_i)=\frac{1}{s}\sum_{x=i}^{s}\frac{1}{x}. \tag{2.4}$$

The proof is as follows (Whitworth, 1934).

Let the lengths of the segments, ranked from largest to smallest, be $y_1, y_2, \ldots, y_s$. Let d_r be the amount by which the length of the rth segment exceeds that of the $(r+1)$th. That is, $d_r = y_r - y_{r+1}$. Then the length of the original stick (unity) can be expressed as the sum of y_s and the d's by the relation

$$1 = sy_s + (s-1)d_{s-1} + \cdots + 2d_2 + d_1.$$

Since the breaks in the stick are located at random, the expected values of the components of this sum are independent and hence are all equal to one another and to $1/s$. That is,

$$E(sy_s)=\frac{1}{s} \qquad \text{whence} \qquad E(y_s)=\frac{1}{s^2}$$

and

$$E(rd_r)=\frac{1}{s} \qquad \text{whence} \qquad E(d_r)=\frac{1}{rs} \qquad \text{for} \quad r=s-1, s-2, \ldots, 1.$$

Therefore

$$E(y_{s-1}) = E(y_s) + E(d_{s-1}) = \frac{1}{s}\left(\frac{1}{s}+\frac{1}{s-1}\right)$$

$$E(y_{s-2}) = E(y_s) + E(d_{s-1}) + E(d_{s-2}) = \frac{1}{s}\left(\frac{1}{s}+\frac{1}{s-1}+\frac{1}{s-2}\right)$$

$$\cdots\cdots$$

and, in general

$$E(y_i)=\frac{1}{s}\sum_{x=i}^{s}\frac{1}{x}.$$

The values of $E(y_i)$ for $i=1, 2, \ldots, s$ constitute the model's expected ranked-abundance list. Now let us derive the corresponding species-abundance distribution by the same method as was used for the niche preemption model in Section 2.2 (and see May, 1975). As before we shall write $\Psi(y)$ and $\psi(y)$ respectively for the cdf and pdf of y, the species' sizes, and shall approximate the true, discrete functions with continuous functions.

Since s is now assumed to be large we may put

$$\frac{1}{s}\sum_{x=i}^{s}\frac{1}{x}\simeq\frac{1}{s}\int_{i}^{s}\frac{dx}{x}=\frac{1}{s}\ln\frac{s}{i}.$$

Then the equation corresponding to (2.1) of the preceding section is

$$y_i=\frac{1}{s}\ln\frac{s}{i} \tag{2.5}$$

whence

$$i=s\exp\left[-sy_i\right] \tag{2.6}$$

which corresponds to (2.2). Putting $1-\Psi(y)\propto i$ as before, we find that

$$\psi(y)=\Psi'(y)=Cs^2\exp\left[-sy\right],\qquad 0<y<\infty \tag{2.7}$$

where C is a constant of proportionality. Putting $\int_0^\infty \psi(y)\,dy=1$ shows that $C=1/s$ and therefore, that $\psi(y)=s\exp[-sy]$ which is the negative exponential distribution with parameter s.

Plots of the cdf and pdf of the distribution are shown in Figure 2.2*a* and *b* respectively. The figure is comparable in all respects with Figure 2.1.

An explanation should be given as to why the range of values specified for y are not the same in (2.3) and (2.7). The reason is as follows. Recall that both equations give continuous functions that are merely approximations to the true discrete distributions. To ensure that the area under the curve in Figure 2.1*b* shall be finite, it is necessary to specify that $y_{min}=(1-k)^{s-1}$ is the smallest, and $y_{max}=k$ is the largest of the species sizes; for if we put $\lim_{s\to\infty} y_{min}=0$ and $\lim_{s\to\infty} y_{max}=\infty$, the area under the curve becomes infinite. In the case of the less skewed curve of Figure 2.2*b* [defined by (2.7)], the area under it is finite even when $0<y<\infty$.

Testing the goodness of fit of the model to field data will be discussed in Section 4.2. Some examples of natural communities that *seem* to be well fitted have been given by King (1964). He found close correspondence between predicted and observed species-abundances in samples of seven species of minnows (Cyprinidae) collected from freshwater streams, for nine species of ophiuroids in an abandoned intertidal quarry in Eniwetok Island, and for five species of *Morula* (a littoral gastropod) in a small area of reef flat off Eniwetok Island.

It is widely believed that good fits of the model are obtained only when the species defined as constituting the community to be tested are few in number and taxonomically close; and when the area (or volume) defined as containing the community is small and very homogeneous. Except in communities thus restrictively defined, species are usually much less evenly represented than the model appears to predict. However, in Section 4.2 we shall see that there is, in fact, no foundation whatsoever for these arguments and conclusions.

The broken stick model is a resource-apportioning model whose predictions are duplicated by a statistical model (see Section 2.1). As shown above, the ranked-abundance list predicted by the broken stick line of argument corresponds to a species-abundance distribution of negative exponential form. One might therefore arrive at identical predictions merely by postulating that the abundances of the species in a community

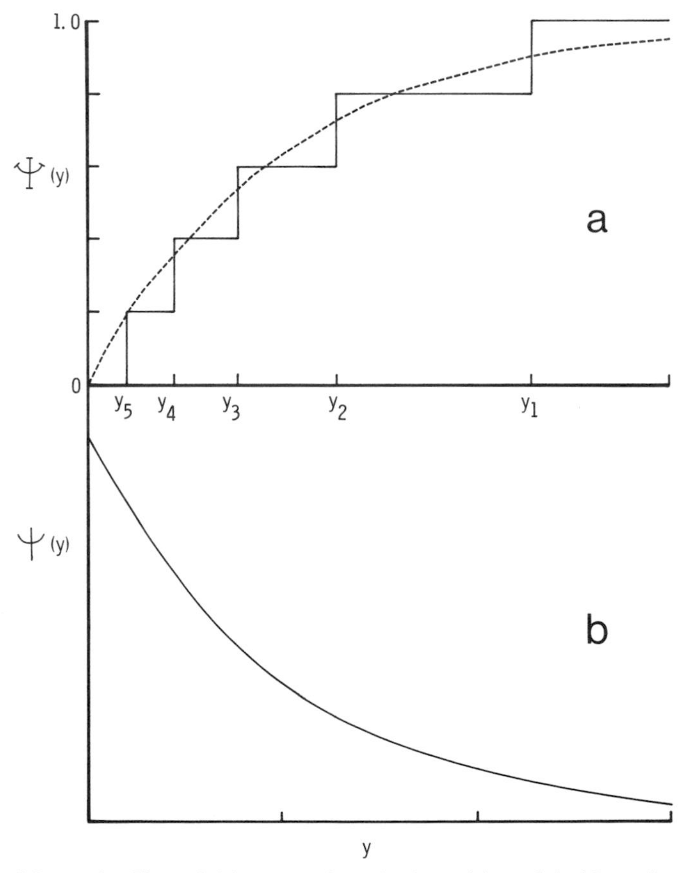

Figure 2.2 As for Figure 2.1 but assuming a broken stick model with $s = 5$.

are proportional to independent, identically distributed variates having a negative exponential distribution (Feller, 1966; Cohen, 1968) and this postulate constitutes a statistical model.

This curious result (curious from the scientific, not the mathematical point of view) should be pondered. The two models (or is it one?) are strikingly different in their "realism." The resource apportioning version has a clearly visualizable, concrete meaning to ecologists; whereas the statistical version is not backed by any intuitively sensible assumptions but only by a theoretical frequency distribution (or probability density function) that might well be described as ignorance in manageable form. Yet the two models yield identical outcomes.

2.4 The Overlapping Niche Model

In the paper in which he first described the broken stick model, MacArthur (1957) also proposed the overlapping niche model; he called these models Hypotheses I and II respectively. Hypothesis II, which we shall now discuss, has received scant attention from ecologists, possibly because the species evenness it predicts is even more pronounced, and hence less realistic, than that of the broken stick model (Pielou and Arnason, 1965). The model does deserve consideration, however.

Hypothesis II as MacArthur described it again likens the environment to a unit "stick" but it supposes that the abundance of each species is proportional to the distance between two points located at random upon it. The species are assumed to be independent of one another and the model differs from the niche preemption and broken stick models in not assuming that some limiting resource must be shared among competitors but rather that each species takes what it needs. From the mathematical point of view, the model is properly a statistical one although originally propounded in terms of resource use.

In contrast to the other two models, this one yields a species-abundance distribution directly and the corresponding ranked-abundance list is derived secondarily. The species-abundance distribution is simply the pdf of the range of a sample of size 2 from the rectangular distribution on the interval [0, 1]. Denoting species "size" (or abundance) by y and its cdf and pdf by $\Psi(y)$ and $\psi(y)$ as before, we therefore have (see, for example, Wilks, 1962)

$$\Psi(y)=(2-y)y$$

and

$$\psi(y)=2(1-y) \qquad 0\leq y\leq 1. \tag{2.8}$$

The ranked-abundance list for the overlapping niche model was derived by Pielou and Arnason (1965). The expected size of the ith largest species out of a total of s is

$$E(y_i)=1-\frac{s!}{(i-1)!}\frac{\Gamma(i-0.5)}{\Gamma(s+1.5)}.$$

Successive values of $E(y_i)$ are most easily obtained (in the order smallest to largest) from the recurrence relation

$$1-E(y_i)=\frac{2i}{2i+1}[1-E(y_{i+1})]$$

by putting $E(y_{s+1})=0$ and then allowing i to take the values $(s-1)$, $(s-2), \ldots, 1$.

Figure 2.3 compares the three models we have now considered; (*a*) shows their pdf's and (*b*) shows plots of $\log E(y_i)$ versus i. It is apparent at a glance that the niche preemption model predicts the lowest evenness of species representation and the overlapping niche model predicts the highest; or in other words, that the dominance of the leading species is most marked in the niche preemption model and least in the overlapping niche model.

In judging which (if any) of the models we have discussed thus far could reasonably account for the species abundance relations of a natural community, we can sometimes do more than compare observed and predicted species-abundance distributions or observed and predicted ranked-abundance lists. It may be possible to observe directly whether or not there is any overlap among the species in the sharing of a resource or of the habitat space.

Observations have often been made on the diets of sympatric congeners to determine whether they eat the same food. For example Ashmole (1968), studying the terns of Christmas Island and the accipiters of North America, found that for both groups of birds many species of prey were common to several members of the group; the same was found to be true of the sandpipers and other insectivorous birds breeding on northern Alaskan tundra (Holmes and Pitelka, 1968). Therefore if, as seems probable, food is the limiting resource for the members of these bird communities, one would not try to fit a true resource apportioning model (such as the niche preemption or broken stick models) to the field observations; of the models discussed thus far, only the overlapping niche model would be appropriate.

A similar judgment can be made when the object of study is a community of plants growing in an environmental gradient and exhibiting zonation as a result. Familiar examples of such zoned communities are

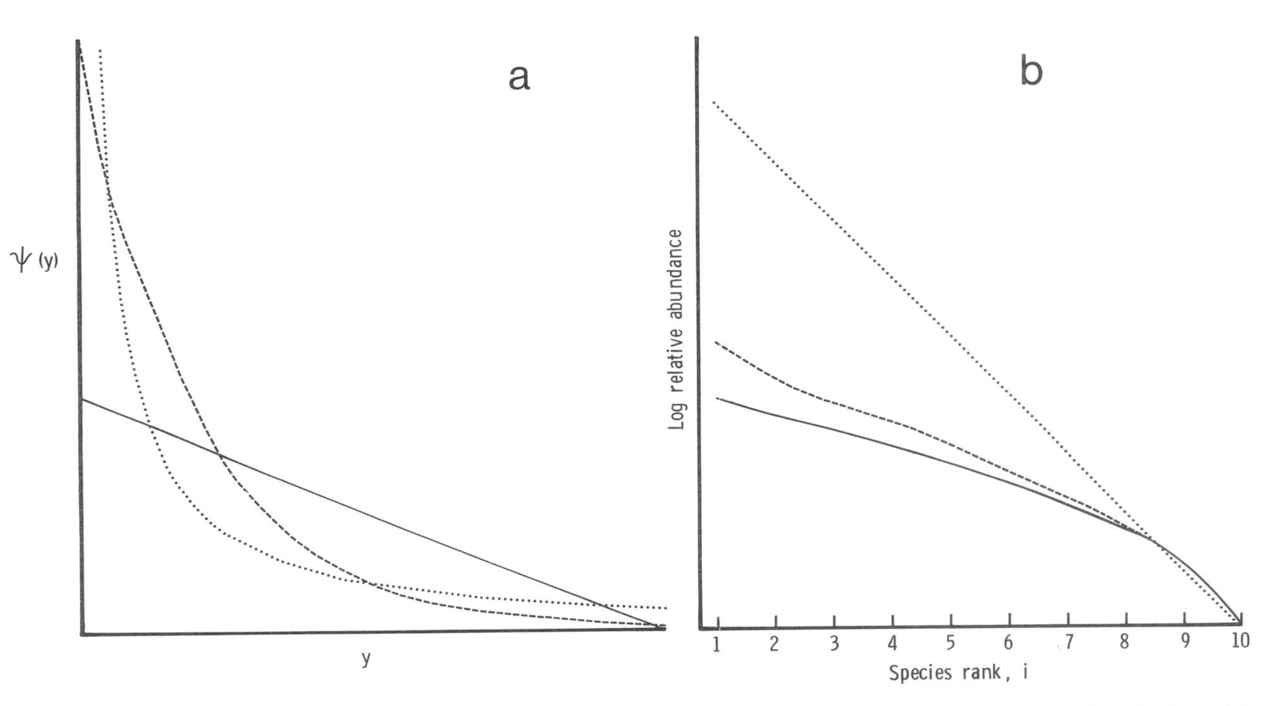

Figure 2.3 (*a*) Plots of $\psi(y)$ versus y; (*b*) plots of log $E(y_i)$ versus *i*. Solid lines: overlapping niche model. Dashed lines: broken stick model. Dotted lines: niche preemption model with $k = 0.5$. For all three models, $s = 10$.

mountain vegetation, salt marshes, the intertidal algae of rocky shores, and the concentric zones of hydrophytes surrounding freshwater ponds. One has only to look to see whether the zones (and, by extension, the "niches") overlap. The model to try can be chosen accordingly: a resource-apportioning model would be appropriate only if the niches were nonoverlapping. Measurement of species' sizes (i.e., abundances) is also unusually straightforward in a zoned community; one has only to measure zone widths.

An additional virtue of zoned communities as objects of study is that numerous replicate measurements of the widths of the zones (and hence of the sizes of the niches) can be had by observing a number of separate transects laid across the zones. Thus suppose the zones overlapped and one wished to test the fit of the overlapping niche model. Measuring zone widths, y, across a transect of length T we find, from (2.8), that

$$E(y)=\frac{T}{3}, \qquad \text{Var}\,(y)=\frac{T^2}{18}$$

and hence that

$$\frac{\text{Var}\,(y)}{E(y)}=\frac{T}{6}.$$

If several transects (sufficiently widely spaced to be independent) are examined, one obtains several independent observations of $\bar{y}$, var (y) and their ratio to compare with expectation. According as var $(y)/\bar{y}$ tends to exceed or fall short of its expected value of $T/6$ one can infer that the species abundances are less or more even than the overlapping niche model predicts.

Figure 2.4 shows an example. The data consisted of observed values of the zone widths of the dominant species* in two Nova Scotia salt marshes in which neighboring zones overlapped strongly. Each dot in the scatter diagram shows the relationship between var $(y)/\bar{y}$ and transect length, T, for a single transect; (the contrasted symbols distinguish data from the two marshes). The transects were rows of contiguous meter-long quadrats and the width of a species' zone was taken to be the number of quadrats between and including those containing the first and last occurrences of the species. Zone width was thus a discrete variate. It may be shown (Pielou,

* The species that dominate the successive zones. From landward to seaward these were: In Marsh 1: *Ligusticum scothicum*, *Solidago sempervirens*, *Juncus Gerardi*, *Carex paleacea*, and *Spartina patens*. In Marsh 2: *S. sempervirens*, *Potentilla anserina*, *J. Gerardi*, *Atriplex patula*, *Glaux maritima*, and *Plantago maritima*. To avoid "end effects" the first and last zones in both marshes (dominated respectively by *Juncus balticus* and *Spartina alterniflora*) were disregarded.

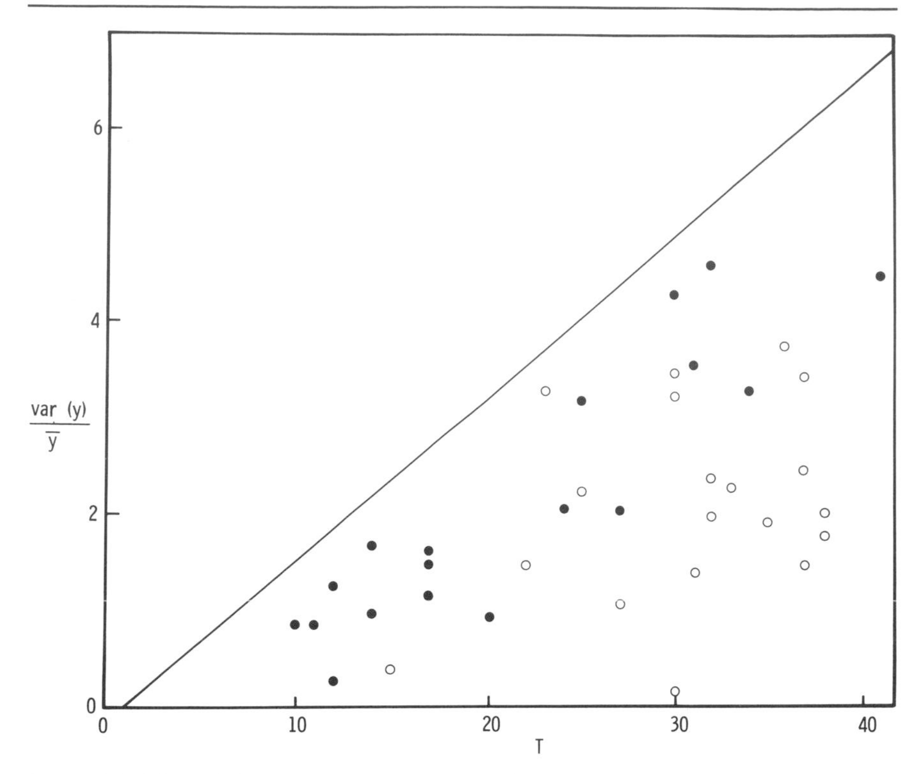

Figure 2.4 The relation between var (y)/ȳ and T for 40 transects across Nova Scotia salt marshes (19 in one marsh and 21 in another). The line shows the relation predicted by the overlapping niche model. (Redrawn from Pielou, 1975.)

1975) that when T and the y's are measured in this way $E(y)=(T+2)/3$, $\text{Var}(y)=(T-1)(T+2)/18$ and thus $\text{Var}(y)/E(y)=(T-1)/6$, which is the line shown in the graph. The result leaves no doubt that the locations of the landward and seaward ends of the zones of the dominant species cannot be regarded as random; that is, their locations could not be simulated by placing pairs of points at random on a stick as the overlapping niche model postulates. As is clear from Figure 2.4, the zone widths were consistently less variable (equivalently, more even) than this model predicts.

Chapter 3

Species Abundance Distributions II

3.1 Introduction

In this chapter we consider three species-abundance distributions derived from "statistical models" (see Section 2.1). The hypotheses constituting each model are assumptions about probability distributions. For example, as we shall see in Section 3.2, the two hypotheses that give rise to the negative binomial distribution are that the numbers of individuals in the several species in a sample are Poisson variates, and that these have parameters that are themselves random variates from a gamma distribution. Details are given below but the point to notice here is that the model is couched entirely in the language of mathematical statistics. It says nothing about the ecology of the species, their environmental tolerances, and whether (and if so how) their abundances are limited by shortages of resources and by competition.

Therefore if a model of this kind fits empirical data, the only conclusion that can be drawn with any confidence is that the probability distributions underlying the model have indeed the forms postulated, at least approximately. But this conclusion is not any less mathematical, or any more ecological, than the assertion that the model itself fits the data. When a statistical model is tested and accepted, one has done no more than accept, as a substitute for a single fairly complicated model, a suite of simpler submodels or hypotheses. The last step, that or arguing back still farther from acceptable statistical hypotheses to acceptable ecological hypotheses, still remains to be done; it is usually the hardest step in an investigation of species abundance relations and one may not succeed at it (nobody has yet), but one should not overlook its existence.

In the light of the foregoing it may well be asked: is anything to be gained by fitting these models to empirical data? I believe there is; all ecological studies that strive for generality are part of a search for repeated, recognizable and, finally, explicable, patterns in the behavior of natural systems. Therefore any uniformities that may exist, such as consistencies in the forms of species-abundance distributions, are worth finding even if for the time being they remain inexplicable. Further, the fitting of statistical distributions to empirical data leads to economy of description; a large mass of data can be summarized by naming the distribution that fits it and giving the estimated parameters of the distribution. This obviously facilitates comparisons among different communities. It would also be worthwhile to discover how the parameters of a model behave when a small community fitted by the model is progressively enlarged in two independent ways: first, in terms of the taxocene whose members are defined as constituting it; second, and independently, in terms of the area (or volume) defined as containing the community. Some interesting regularities might be found.

The distributions described below have, for the most part, been fitted to data obtained by sampling very large communities; [for example, to a sample of collembolans from soil, with over 11,000 individuals in 27 species (Brian, 1953)]. Communities so large that they must be sampled rather than censused often contain an unknown number of species. The number found in a sample, say s, may fall considerably short of the total, s^*, and estimating s^* is often one of the objectives of fitting a theoretical species-abundance distribution to the data.

The larger the sample the greater the information it provides, but it should be noticed that when sample size is increased two things happen: N, the number of individuals in the sample is, of course, increased; and s, the number of species in the sample, may or may not be augmented according as the added individuals are members of "new" species (new to the sample, that is) or merely additional representatives of "old" ones. It should be recalled that the variate whose distribution is being investigated is "species size" (synonymously, species-abundance) and usually (though not necessarily always) the size of a species is measured by the number of individuals it contains. Adding extra individuals to a sample is therefore not the same as adding extra variate values unless the added individuals represent new species. Those of the added individuals that belong to old species cause changes in the observed sizes of the species they belong to; this necessitates making changes in variate values already listed. The $s^* - s$ species that are *un*represented in the sample may be thought of as species whose *observed* sizes are zero (though their true sizes are assumed to be nonzero); or, alternatively, as the possessors of sizes that have not been observed and recorded.

3.2 The Truncated Negative Binomial Distribution

The negative binomial distribution is not often fitted to species-abundance data (but see Brian, 1953). However, since the much-used logseries distribution (see Section 3.3) is a limiting form of the negative binomial, we discuss the latter first.

It should be noticed that in what follows the word *sample* connotes the whole sample taken from a community. If sampling is carried out (as it often is) by observing a number of independent sampling units such as quadrats, then the sample consists of all sampling units combined (cf Section 1.5).

Suppose there are a total of s^* species in a community; suppose also that the number of individuals of the jth species in the sample is a Poisson variate with parameter λ_j and assume that the different values of λ_j constitute s^* independent random variates from a two-parameter gamma distribution (equivalently, a type III distribution). That is, the pdf of the λ's is

$$f(\lambda) = \frac{p^{-k}\lambda^{k-1}\exp[-\lambda/p]}{\Gamma(k)}, \qquad 0<\lambda<\infty \tag{3.1}$$

where the positive constants p and k are the parameters of the gamma distribution.

Then q_r, the probability that any species will be represented in the sample by r individuals is

$$\begin{aligned} q_r &= \int_0^\infty \frac{\lambda^r e^{-\lambda}}{r!} f(\lambda)\, d\lambda \\ &= \frac{p^{-k}}{r!\,\Gamma(k)} \int_0^\infty \lambda^{r+k-1} \exp\left[-\lambda\left(\frac{p+1}{p}\right)\right] d\lambda \\ &= \frac{\Gamma(k+r)}{r!\,\Gamma(k)} \left(\frac{p}{1+p}\right)^r \left(\frac{1}{1+p}\right)^k \qquad \text{for} \quad r=0, 1, 2, \ldots . \end{aligned} \tag{3.2}$$

This is the general term of the negative binomial distribution. It implies that the probability that a species will contain 0 individuals is $q_0 = (1+p)^{-k} > 0$. But "empty" (i.e., unrepresented) species are unobservable. Thus the proportion of the observed species with r individuals is predicted to be

$$q'_r = \frac{q_r}{1-q_0} = \frac{\Gamma(k+r)}{r!\,\Gamma(k)}\left(\frac{p}{1+p}\right)^r \frac{1}{(1+p)^k - 1} \qquad \text{for} \quad r = 1, 2, \ldots. \tag{3.3}$$

This is the general term of the zero-truncated negative binomial distribution. Its first and second moments about the origin are, respectively,

$$\mu'_1 = \frac{kp}{1-(1+p)^{-k}} \qquad \text{and} \qquad \mu'_2 = [1+p(1+k)]\mu'_1.$$

Then the mean and variance are

$$E(r) = \mu'_1 \qquad \text{and} \qquad \text{var}\,(r) = \mu'_2 - \mu_1'^2.$$

The parameters p and k may be estimated from the moments of the observed species-abundance distribution. This is most easily done as follows: (see Table 3.1 for a numerical example in which the fit turned out to be poor).

Let the sample contain N individuals belonging to s species. Let f_r be the number of species containing r individuals.

Then $\sum f_r = s$; $\sum rf_r = N$; and $\bar{r} = N/s$ is the mean number of individuals per species, or the observed mean of the distribution to be fitted. The second moment about the origin of the observed distribution is $m'_2 = \sum r^2 f_r / \sum f_r$. Equating observed and expected moments yields

$$\bar{r} = \frac{kp}{1-(1+p)^{-k}}; \qquad \frac{m'_2}{\bar{r}} = p(1+k) + 1 \tag{3.4}$$

which can be solved by successive approximation.

To calculate the expected values of f_r, namely

$$E(f_r) = sq'_r \qquad \text{for} \quad r = 1, 2, \ldots$$

one may start by putting $r = 1$ in (3.3) to give

$$q'_1 = k\left(\frac{p}{1+p}\right)\left(\frac{1}{(1+p)^k - 1}\right)$$

and then use the recurrence relation

$$q'_{r+1} = \left(\frac{k+r}{1+r}\right)\left(\frac{p}{1+p}\right)q'_r \qquad \text{for} \qquad r = 2, 3, \ldots.$$

Species-abundance distributions are apt to have long tails owing to the sporadic occurrence of very large r values. Therefore logarithmic grouping of the frequencies is usually desirable. The customary method of grouping is that first proposed by Preston (1948). The group boundaries are $1, 2, 4, 8, \ldots, 2^x, \ldots$ so that the logs (to base 2) of the boundaries are the arithmetic series $0, 1, 2, 3, \ldots$. This method of grouping has the disadvantage that the group boundaries are integers; this necessitates splitting some of the observed frequencies (those for variate values that are powers of 2) between two adjacent groups. Using values of $\ln r$ as group boundaries, as has been done in Table 3.1, is therefore more convenient.

To estimate expected frequencies in higher groups, the best method (Brian, 1953) is to obtain spot values of q_r' for group boundary values of r and then to approximate group frequency by assuming that the expected frequency in the group whose boundaries are, say, $r=i$ and $r=j$ (with $i<j$) is $s(j-i)(q_i'+q_j')/2$. The spot values of q_r' are calculated from (3.3); the required gamma functions may be found from tables such as those in Khamis and Rudert (1965).

When the zero-truncated negative binomial distribution does fit the observations, estimation of the unknown s^* becomes possible. Equating the "observed" and expected frequencies of "empty" (or unobserved) species shows that

$$E(s^*-s)=s^*(1+p)^{-k}$$

from (3.2). Hence as an estimator of s^* we may take

$$\tilde{s}^*=\frac{s}{1-(1+\tilde{p})^{-\tilde{k}}} \tag{3.5}$$

where $\tilde{p}$ and $\tilde{k}$ are the estimates of the parameters p and k obtained by equating observed and expected moments. The sampling variances of $\tilde{p}$, $\tilde{k}$, and $\tilde{s}^*$ are unknown.

Of the parameters p and k, only k has ecological import; p can be expressed as a function of N, k, and s^*, as follows. From (3.4)

$$\bar{r}=\frac{N}{s}=\frac{kp}{1-q_0}=\frac{kps^*}{s} \quad \text{whence} \quad p=\frac{N}{ks^*}. \tag{3.6}$$

(We drop the distinction between parameter values and their estimators.) Thus if N is increased, p is increased. The reason for regarding p rather than k as arbitrary is explained below. A summary description of the species abundance relations of the sample (and by inference, of the whole

Table 3.1 The species-abundance distribution of 72 species of insects and spiders emerging from 257 sporophores of the birch bracket fungus *Polyporus betulinus* collected in New Brunswick, Canada[a]

The raw data:

r	f_r	r	f_r	r	f_r	r	f_r
1	31	10	2	32	1	87	1
2	3	12	1	33	1	91	1
3	2	15	1	40	1	97	1
4	5	16	1	48	1	98	1
5	1	17	1	49	1	114	1
6	2	18	1	67	1	142	1
7	3	20	1	71	1	196	1
8	1	25	1	84	1		

The constants and parameter estimates:

$s=\sum f_r=72$; $N=\sum rf_r=1501$; $\bar{r}=20.847$;

$m_2'=1866.0417$; $\text{var}(r)=1431.435$; $\frac{m_2'}{\bar{r}}-1=88.5103$;

hence $\tilde{k}=0.0398$ and $\tilde{p}=85.1257$.

Comparison of the observed and expected (zero-truncated negative binomial) distributions:

		Frequencies	
$\ln r$	r	Observed, O	Expected, E
0–1	1, 2	34	22.10
1–2	3–7	13	16.50
2–3	8–20	9	14.38
3–4	21–54	6	11.47
>4	>54	10	7.55

$\sum\frac{(O-E)^2}{E}=12.57$; $P(\chi^2\geq 12.57 \mid \nu=2)<0.002$.

[a] From unpublished data and see Pielou and Verma (1968).

community) is thus given by two numbers: s^*, the total number of species; and k, which measures the shape of the species-abundance distribution. The more prolonged the tail of the distribution (equivalently, the greater the proportion of rare species), the smaller the value of k.

The way in which s increases with N can also be predicted. The resulting equation is that of the so-called "collector's curve" relating the number of species collected with collecting effort. Clearly, from (3.5) and (3.6),

$$s = s^*\left[1-\left(1+\frac{N}{ks^*}\right)^{-k}\right]. \tag{3.7}$$

As one would expect, the curve approaches s^* asymptotically.

This equation is rarely of more than academic interest. To augment a sample by increasing the area examined or by prolonging the operating time of a collecting device such as an insect light-trap has the effect of redefining the community under study by redefining its boundaries; it does *not* have the effect of increasing the intensity of sampling of the original community. Such an increase can only be achieved when the original sample consisted of sampling units located at random in the community's area; one can then increase the sample by collecting more sampling units in the same manner. Even then the randomness of the collecting method relates to the space being sampled and not to the species occupying the space (cf Section 1.5) which may occur in large patches. Thus although collector's curves are qualitatively interesting, the way in which they are constructed seldom justifies the fitting of a precise theoretical curve.

Up to this point we have assumed species abundances to have been measured by counts of individuals. Thus the observed variates were discrete and a discrete theoretical distribution was fitted to the empirical data. But counting individuals is not the only way, and often is not the natural way, of estimating species "size." For many kinds of communities it is more appropriate to use continuously varying quantities like weight or volume as measures of species quantity; for instance one may determine fresh weights or dry weights, or (for vegetatively reproducing plants or colonial invertebrates) "cover" (cf Section 1.1). The observed variate is then continuous and a fitted theoretical distribution must be continuous likewise.

This suggests fitting a two-parameter gamma distribution to data of this kind. It will be recalled that we derived the truncated negative binomial as the discrete analog of the gamma distribution, for use when the observed variate was discrete. Obviously the distribution to try when the variate is continuous is the gamma distribution itself. That is, we assume λ_j, the quantity of species j, to be a random variate from the distribution with pdf

$$f(\lambda)=\frac{p^{-k}\lambda^{k-1}\exp[-\lambda/p]}{\Gamma(k)}, \qquad 0<\lambda<\infty$$

as in (3.1).

The mean and variance of the distribution are

$$E(\lambda)=kp \qquad \text{and} \qquad \mathrm{Var}(\lambda)=kp^2. \tag{3.8}$$

Therefore the moment estimators of the parameters are

$$p=\frac{\mathrm{var}\,(\lambda)}{\bar{\lambda}} \qquad \text{and} \qquad k=\frac{\bar{\lambda}^2}{\mathrm{var}\,(\lambda)}.$$

The expected proportion of species abundances in the range (λ_1, λ_2), say, is

$$\Pr\{\lambda_1<\lambda<\lambda_2\}=F(\lambda_2)-F(\lambda_1)$$

where $F(\lambda)$ is the cdf of the distribution, namely

$$F(\lambda)=\frac{p^{-k}}{\Gamma(k)}\int_0^\lambda x^{k-1}\exp\left[-\frac{x}{p}\right]dx.$$

To evaluate this expression, make the substitution $y=2x/p$. Then

$$F(\lambda)=\frac{1}{2^k\Gamma(k)}\int_0^{2\lambda/p} y^{k-1}e^{-y/2}\,dy. \tag{3.9}$$

Published tables of this function (which is the probability integral of the χ^2 distribution with $2k$ degrees of freedom for $\chi^2=2\lambda/p$) are available (Khamis and Rudert, 1965). An alternative method for evaluating $F(\lambda)$ that does not entail interpolation among tabulated values is possible when $2\lambda/p\leq 1$. One can then use the relation (Pearson and Hartley, 1954)

$$F(\lambda)=\exp\left[-\frac{\lambda}{p}\right]\sum_{j=0}^{\infty}\left(\frac{\lambda}{p}\right)^{k+j}\frac{1}{\Gamma(k+j+1)}. \tag{3.10}$$

When $2\lambda/p$ is small the sum converges rapidly. The gamma function required to start the summation, $\Gamma(k+1)$, is obtainable from the Khamis and Rudert tables.

A numerical example is given in Table 3.2. The data (given in full in Pielou, 1966*b*) are a list of the fresh weights in grams of $s=62$ species of herbaceous plants collected from sampling units totaling 100 sq m in area; these were located at random in a community defined as occupying 3000 sq m of deciduous woodland. The parameters of the fitted distribution were estimated from the moments of the observed distribution and (3.10) was used to calculate the needed values of $F(\lambda)$. The expected number of species for which the variate lay in the interval (λ_1, λ_2) is then given by $s[F(\lambda_2)-F(\lambda_1)]$.

As with the truncated negative binomial distribution, so also with the gamma distribution: the parameter k determines the shape of the distribution whereas p is arbitrary. Indeed, if we were to divide all observations by p (which merely amounts to changing the units in which the variate is measured), we should have, from (3.8),

$$E(\lambda)=\mathrm{Var}\,(\lambda)=k$$

Table 3.2 The species-abundance distribution of 62 species of herbaceous plants in an area of woodland. The data are a list of the fresh weights (in grams) of the species in a sample (Pielou, 1966*b*)

Constants and parameter estimates:

$s=62$; $\bar{\lambda}=161.49677$ g var $(\lambda)=75015.27613$.
Hence $\tilde{p}=464.50$; $\tilde{k}=0.34768$.

Comparison of the observed and expected (gamma) distributions

λ Species Weight grams	Number of species	
	Observed, *O*	Expected, *E*
0–2.9	7	12.02
3.0–5.9	5	3.25
6.0–	8	4.10
12.0–	3	5.12
24.0–	5	6.27
48.0–	11	7.40
96.0–	7	8.08
192.0–	8	7.35
≥384.0	8	8.41
	62	62.00

$\sum\frac{(O-E)^2}{E}=9.857$; $P(\chi^2\geq 9.857 \mid \nu=6)=0.13$.

and

$$f(\lambda)=\frac{\lambda^{k-1}e^{-\lambda}}{\Gamma(k)} \qquad \text{for} \quad 0<\lambda<\infty. \tag{3.11}$$

This is the customary form of the pdf of the gamma distribution with one parameter, k.

There has been much debate among ecologists as to whether, in large communities, rare species are always (or nearly always) more numerous than common ones, in other words whether, when the observed measure of species abundance is discrete, it is true in general that $f_r>f_{r+1}$ for all r as is usually found to be the case. Opinion is divided as to whether a modal value of f_r for some $r>1$ is commonplace or exceptional (and then perhaps due to sampling error). It is therefore interesting to consider the shape of the gamma distribution for different values of k; the conclusions apply also

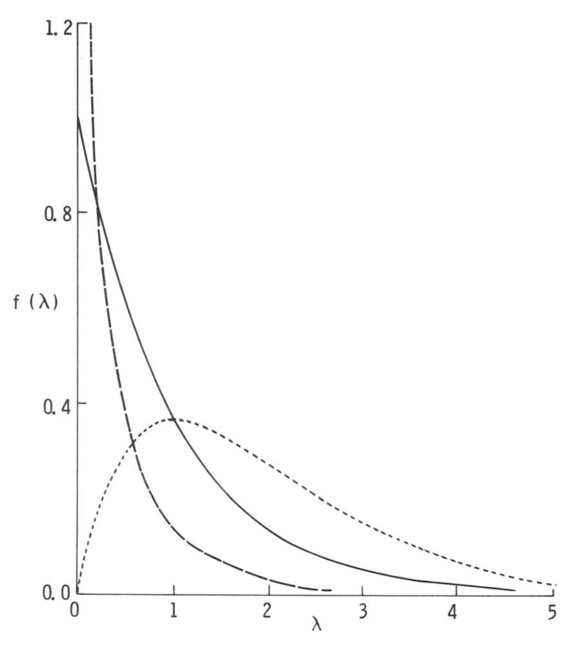

Figure 3.1 The pdf of the gamma distribution, $f(\lambda)=\lambda^{k-1}e^{-\lambda}/\Gamma(k)$ with $k=2$ (dotted line); $k=1$ (solid line); and $k=0.348$ (dashed line) which is the estimated k for the distribution fitted in Table 3.2.

to the truncated negative binomial distribution. Differentiating (3.11) shows that $f'(\lambda)=0$ when $\lambda=k-1$. Since we must have $\lambda>0$, it follows that $f(\lambda)$ will have a mode to the right of the origin only if $k>1$, in which case also, $f(0)=0$.

If $k\leq 1$, $f'(\lambda)<0$ for all $\lambda>0$. Further, if $k<1$, $f(0)=\infty$; and if $k=1$, $f(\lambda)=e^{-\lambda}$ and $f(0)=1$. Examples of gamma distribution pdf's are shown in Figure 3.1.

Absence of a mode in a fitted theoretical curve cannot of course be taken to imply absence of a mode in the observations themselves. However, Brian (1953) remarked on the fact that for all the negative binomial distributions he fitted to actual data, k was consistently less than unity. He also pointed out that when variate values are grouped logarithmically and the frequencies per group shown as a histogram, presence of a mode in the histogram does not necessarily imply presence of a mode in the pdf; it may be an artefact. The possibly misleading effect of logarithmic grouping can be demonstrated by comparing plots of the cdf of the gamma distribution with the abscissa arithmetically scaled and logarithmically scaled, both with

$k<1$ (see Figure 3.2). When λ is scaled arithmetically, the $F(\lambda)$ curve (solid line) exhibits no point of inflection (its second derivative is everywhere positive) which implies a monotonically decreasing pdf. The inflection apparent in the $F(\lambda)$ curve when λ is scaled logarithmically (dotted line) results from the distortion of the scale on the abscissa. It does *not* imply a mode in the pdf.

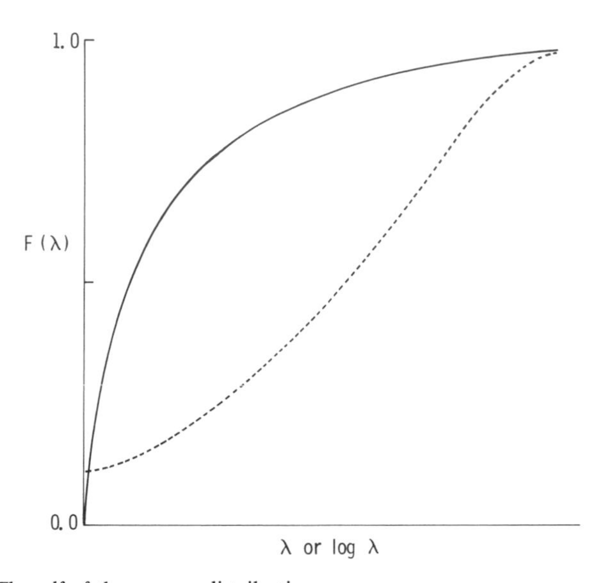

Figure 3.2 The cdf of the gamma distribution

$$F(\lambda) = [1/\Gamma(k)] \int_0^\lambda x^{k-1} e^{-x} \, dx$$

with $k = 0.5$. The solid line shows the cdf when the scaling of the abscissa is arithmetic, and the dotted line when the scaling is logarithmic.

3.3 The Logseries Distribution

According to the logseries distribution (otherwise called the logarithmic series distribution, or the logarithmic distribution) the frequency of species with r individuals in a community is

$$f_r = \frac{\alpha X^r}{r} \quad \text{for} \quad r = 1, 2, \ldots \tag{3.12}$$

where $\alpha > 0$ and $0 < X < 1$ are constants.

The distribution was first derived, nonrigorously, by allowing k to tend to zero in the expression for q_r', the general term of the truncated negative binomial distribution. The idea of allowing k to tend to zero was first suggested by Fisher (in Fisher, Corbet and Williams, 1943) and was prompted by the fact that many of the bodies of data he analyzed yielded values of k so small as to be "indistinguishable from zero" although, strictly speaking, k must be finite.

Recall (3.3), namely

$$q'_r = \frac{\Gamma(k+r)}{r!\,\Gamma(k)}\left(\frac{p}{1+p}\right)^r \frac{1}{(1+p)^k - 1}.$$

Putting $p = X/(1-X)$ and collecting terms independent or r into a single constant, C, shows that

$$q'_r = C\frac{\Gamma(k+r)}{r!}X^r.$$

Then $\lim_{k\to 0} q'_r = \pi_r$, say, is

$$\pi_r = \lim_{k\to 0} C\frac{\Gamma(k+r)}{r!}X^r = \frac{X^r}{r}\lim_{k\to 0} C.$$

As Holgate (1969) has shown, it is not mathematically permissible to put $k=0$ in the pdf of the gamma distribution (on which the formula for q'_r is based). We can, however, complete the specification of π_r by noting that if $\pi_r = \gamma X^r/r$ (where γ is a constant of proportionality) then, since we must have $\sum_{r=1}^{\infty} \pi_r = 1$, it follows that $\gamma = -1/\ln(1-X)$. Observe that since γ is expressible as a function of X (or X of γ) the distribution has only one parameter. However, as with the gamma distribution (see Section 3.2) it is convenient to retain both γ and X to facilitate algebraic manipulations, while bearing in mind that X $(= p/(1+p))$ is merely a scale factor whereas γ depends on the shape of the distribution.

Since X is arbitrary, two parameters suffice to describe the species-abundance distribution of a sample; these are s, the observed number of species, and γ which depends on the shape of the distribution. Instead of s and γ however, it is customary to use s and $\alpha = s\gamma$; the longer the tail of the distribution the greater the value of α and this parameter therefore serves as an index of diversity. It was so used by Williams (see Fisher, Corbet, and Williams, 1943) who was the first to introduce the concept of community diversity and to propose a method for measuring it.

Notice that it is s (the number of species in the sample), not s^* (the estimated number in the whole community) that, with α, is used to describe the population. This is because in assuming that a sample comes from a community whose species abundances have a logseries distribution, one is also assuming that s^* is "infinite" (in effect, indefinitely large). To see this, consider the expected form of the "collector's curve" which is derived as follows. Since

$$s=\sum_{r=1}^{\infty} f_r = s\sum \pi_r = s\gamma \sum \frac{X^r}{r} = -\alpha \ln(1-X) \tag{3.13}$$

and

$$N=\sum_{r=1}^{\infty} rf_r = s\sum r\pi_r = s\gamma \sum X^r = \frac{\alpha X}{1-X}, \tag{3.14}$$

elimination of X shows that

$$s=\alpha \ln\left(1+\frac{N}{\alpha}\right). \tag{3.15}$$

Thus the number of species in a sample is assumed to increase indefinitely as sample size, N, is increased. In contrast to the collector's curve for the truncated negative binomial distribution [see (3.7)], (3.15) has no upper limit which the curve approaches asymptotically. Indeed, for large N, s increases almost linearly with $\ln N$. This is often found to be approximately true in practice. Ecologists who try to draw up exhaustive catalogs of the species in large communities are accustomed to finding that doubling the size of the sample tallied does not (in practice) lead merely to a doubling of the numbers of all the species already encountered, which would leave the observed s and the relative species-abundances unaltered; it usually brings in also, in small numbers, a few "new" species. This is the reason why it is desirable that the diversity index of a finite collection should be dependent on the collection's size (cf. Section 1.4).

To fit a logseries to empirical data, it is first necessary to estimate α and X from (3.13) and (3.14) and then to evaluate the terms $f_r=\alpha X^r/r$ for $r=1, 2, \ldots$. A rapidly converging iterative algorithm for obtaining α and X has been given by Birch (1963). Calculating the expected frequencies $\alpha X^r/r$ is straightforward for small values of r. For higher, grouped values, one requires sums of the form $\sum_{r=a}^{b} \alpha X^r/r$, and if $b-a$ is large it is convenient to use the following formula (Bliss, 1965).

$$\sum_{r=a}^{b} f_r \simeq \alpha\left\{\sum_{r=a}^{b}\frac{1}{r}-(b-a+1)(1-X)+\frac{b^{(2)}-(a-1)^{(2)}}{2(2!)}(1-X)^2 - \frac{b^{(3)}-(a-1)^{(3)}}{3(3!)}(1-X)^3+\cdots + (-1)^m\frac{b^{(m)}-(a-1)^{(m)}}{m(m!)}(1-X)^m\right\}$$

where $b^{(m)} = b(b-1)(b-2) \cdots (b-m+1)$. Sufficient accuracy for most purposes is attained by letting $m=3$ or 4.

Table 3.3 shows a logseries fitted to the data (on insects and spiders in bracket fungi) already given in Table 3.1. For comparison, the fitted negative binomial distribution is repeated. The two theoretical distributions are closely similar, as one would expect since the negative binomial parameter k ($\doteq 0.04$) has a low value; both give a very poor fit. The parameters of the logseries distribution, calculated by Birch's algorithm, are estimated by $\hat{\alpha} = 15.768$ and $\hat{X} = 0.9896$.

Anscombe (1950) has shown that an estimator of the sampling variance of $\hat{\alpha}$ is given by

$$\widehat{\text{var}}\,(\hat{\alpha}) \simeq \frac{\hat{\alpha}}{\ln X(1-X)};$$

(a helpful exposition of Anscombe's arguments, with more intermediate steps, has been given by Nelson and David, 1967). Anscombe also shows that the distribution of $\hat{\alpha}$ is asymptotically normal. We can therefore determine a confidence interval for α, as should be done if it is to be used as a diversity index. However, it should *not* be so used unless the observations are well fitted by a logseries distribution. For most purposes, a diversity index whose legitimacy does not depend on the form of the observed distribution is preferable and the indices H' and H discussed in Chapter 1 meet this requirement.

There is no continuous distribution corresponding to the logseries

Table 3.3 Comparison of a fitted logseries distribution with the observed distribution, and the fitted negative binomial distribution, given in Table 3.1

		Expected	
$\ln r$	Observed	Negative binomial	Logseries
0–1	34	22.10	23.32
1–2	13	16.50	16.47
2–3	9	14.38	12.20
3–4	6	11.47	10.76
≥5	10	7.55	9.25
	72	72.00	72.00

distribution in the way that the gamma distribution corresponds to the negative binomial, since for the gamma distribution $k>0$ by definition.

3.4 The Lognormal Distribution

As a species-abundance distribution, the lognormal may be thought of as based on either a "statistical model" or a "resource apportioning model" (see Section 2.1).

Consider it first as the outcome of a statistical model. If we suppose that the number of individuals belonging to a given species in a community results from the combined effect of a large number of mutually independent causes that are multiplicative in their effect, then the abundance of the species is a lognormal variate. To see this, recall the central limit theorem: it states that the sum, say z, of n independent variates with finite means and variances has a distribution that approaches normality asymptotically with increasing n. That is, if $z=\zeta_1+\zeta_2+\cdots+\zeta_n$, in the limit the pdf of z is

$$f(z)=\frac{1}{\sqrt{2\pi V_z}}\exp\left[\frac{-(z-\mu_z)^2}{2V_z}\right], \qquad -\infty<z<\infty$$

where μ_z and V_z are, respectively, the mean and variance of the distribution. Notice that the ζ's need not have the same distribution; to ensure that their sum shall have statistically predictable properties, it is only necessary that there be enough of them.

Now suppose that the variate of interest is not the sum but the product of a very large number of independent factors. This will be the case if we assume the variate to have undergone a sequence of n changes in value, with the magnitude of each successive change being a random proportion of the variate's immediately preceding value. Thus if y_j denotes its magnitude after the jth change, we assume that

$$y_j-y_{j-1}=\varepsilon_j y_{j-1} \tag{3.16}$$

where the ε_j (which may be positive or negative) are random variates that are mutually independent and independent of the y's. Then

$$y_n=(1+\varepsilon_n)(1+\varepsilon_{n-1})\cdots(1+\varepsilon_1)y_0$$

and

$$\ln y_n=\ln y_0+\sum_{j=1}^{n}\ln(1+\varepsilon_j).$$

Assuming $\varepsilon_j \ll 1$ so that $\ln(1+\varepsilon_j)\simeq\varepsilon_j$, we may take

$$\ln y_n=\sum_{j=1}^{n}\varepsilon_j+\text{constant}.$$

Then since, by the central limit theorem, $\lim_{n\to\infty} \sum_{j=1}^{n} \varepsilon_j$ is normally distributed, so is $\lim_{n\to\infty}(\ln y_n)$. This amounts to saying that $\lim_{n\to\infty} y_n = y$, say, is lognormally distributed.

The foregoing argument justifies us in entertaining the hypothesis that the sizes of separate and independent populations of some one species (in separate noninteracting communities) will be random variates from a lognormal distribution, for it is reasonable to assume that (3.16) describes the successive changes in size that such populations undergo. But it does *not* follow (as is sometimes assumed) that the distribution of the population sizes of a number of *different* species, occurring together and forming a single community, must also be lognormal. Therefore the "statistical model" that leads to the lognormal distribution is purely statistical and has no claim to biological realism.

However, the "resource apportioning model" that leads to the lognormal does have a claim (not perhaps a very strong one) to being thought realistic. Conceived of as an elaboration of the "broken stick" model, a simple version of the argument is as follows. Imagine a stick of length L_0 marked at one end. Let the stick be broken, at a randomly chosen point, into two parts. Choose one of the parts at random and break it again. There are now three parts; choose one of the three at random and break it again. And so on. The important point to notice is that at each step every part, regardless of its length, has an equal chance to be chosen for the next break: that is, the probability that a part will be broken is *independent of its length.* After a very large number of breaks has been made, the part with the mark on one end has length $L_n = L_0 \prod_{j=1}^{n} r_j$ where n is the number of breakages this part has undergone and r_j is the length of the marked part as a fraction of its length before the jth breakage. Clearly, L_n is the product of a large number of independent positive variates and is thus a lognormal variate. The same is true of all the parts into which the stick has been broken: their probability distributions are identical. Thus the distribution of the lengths of all the parts tend to lognormal form. The argument also applies if, at each step, the chosen part is broken more than once; the breaks need not even be at random; it is necessary only that the number of pieces into which a part is broken be independent of its length (Aitchison and Brown, 1966; Bulmer, 1974). If we now assume (as in Section 2.3) that the species in a many-species taxocene constituting a community have divided up some limiting resource among themselves in this manner, and that the abundance of each species is proportional to its share of the resource, then the species-abundance distribution is lognormal. That is, the pdf of y, the "size" of the species, is taken to be

$$\phi(y)=\frac{1}{y\sqrt{2\pi V_z}}\exp\left[\frac{-(\ln y-\mu_z)^2}{2V_z}\right], \qquad 0<y<\infty$$

which is the pdf of the lognormal distribution. The mean and variance are

$$\mu_y=\exp\left[\mu_z+\frac{V_z}{2}\right]; \qquad V_y=(\exp[V_z]-1)\exp[2\mu_z+V_z]. \qquad (3.17)$$

Here μ_z and V_z are the moments of z, the normally distributed variate; and μ_y and V_y are the moments of $y=e^z$ (so that $\ln y=z$), the lognormally distributed variate.

The lognormal distribution is continuous by definition, whereas if species-abundances are measured by counting individuals, the observed variate is discrete. One could regard the observed species-abundances as compound Poisson variates with the Poisson parameter, λ, being lognormally distributed. Then the probability, π_r, that a species contains r individuals is

$$\pi_r=\int_0^\infty \frac{e^{-\lambda}\lambda^r}{r!}\phi(\lambda)\,d\lambda$$

$$=\frac{1}{r!\sqrt{2\pi V}}\int_0^1 \frac{1}{\lambda}\exp\left[-\lambda+r\ln\lambda-\frac{(\ln\lambda-\mu)^2}{2V}\right]d\lambda \qquad \text{for} \quad r=0,1,2,\ldots. \qquad (3.18)$$

This distribution is variously called the Poisson lognormal (Holgate, 1969; Bulmer, 1974) or the discrete lognormal (Anscombe, 1950; Pielou, 1969); the former name is more informative. Bulmer (1974) has obtained an approximate formula permitting evaluation of the probabilities π_r for $r\geq 10$ but can find no alternative to numerical integration of the integral in (3.18) for low values of r. The Poisson lognormal is therefore computationally troublesome and its use may introduce no more than a spurious precision since it assumes that all individuals are equal. Use of the ordinary lognormal therefore seems justified and it is, of course, the appropriate distribution to fit when species quantities are measured in terms of a continuous variate such as biomass instead of by counting individuals (cf. Section 3.1).

We now describe the fitting of a lognormal distribution to field data. To illustrate, the distribution will be fitted to the observed species-abundances in one of the experimental diatom communities described by Patrick (1968). The raw data are reproduced here in Table 3.4.

Table 3.4 The species-abundance distribution of 113 species in an experimental community of diatoms[a]

r	f_r	r	f_r	Values of r for which $f_r=1$		
1	16	11	3			
2	10	17	6			
3	5	19	3	6	38	147
4	10	20	4	7	62	184
5	8	21	2	23	67	192
8	2	33	4	24	75	408
9	6	36	2	26	82	640
10	9	272	2	32	110	2960
				37	124	3032

$s=\sum f_r=113 \qquad N=\sum rf_r=9629$

[a] Data from Table 3 (box and slide B-7-2) of Patrick (1968)

The procedure consists in converting the observed variate, r, to logs (we shall put $x = \log_{10} r$) and then fitting a normal distribution to the x's. Recall that we are treating r as a continuous variate. To do this, we substitute, for the discrete values $0, 1, 2, \ldots, r, \ldots,$ the intervals $(0, \frac{1}{2}]$, $(\frac{1}{2}, 1\frac{1}{2}]$, $(1\frac{1}{2}, 2\frac{1}{2}] \cdots (r-\frac{1}{2}, r+\frac{1}{2}] \ldots$. The fitting would then be straightforward if it were not for the fact that the observed distribution is zero-truncated. Since "empty" (unrepresented) species are unobservable, the value $r=0$ is missing from the raw (discrete) data, and the interval $(0, \frac{1}{2}]$ from the continuous representation of the data. Hence the normal distribution to be fitted to the x's is truncated on the left at $x_0=\log_{10} 0.5=-0.30103$.

In recipe form, the procedure is now as follows; (Table 3.5 gives numerical results for the diatom example).

1. Put $\log_{10} r=x$.
2. Obtain the observed mean and variance of x; namely, $\bar{x}=\sum x/s$ and $\sigma^2=\sum (x-\bar{x})^2/s$. ($s$ is the observed number of species.)
3. Calculate $\gamma=\sigma^2/(\bar{x}-x_0)^2$ where $x_0=-0.30103$.
4. From Table 1 in Cohen (1961) obtain the "auxiliary estimation function" $\hat{\theta}$ corresponding to this γ.
5. Obtain estimates $\hat{\mu}_x$ and $\hat{V}_x$ of the mean and variance of x from $\hat{\mu}_x=\bar{x}-\hat{\theta}(\bar{x}-x_0)$; $\hat{V}_x=\sigma^2+\hat{\theta}(\bar{x}-x_0)^2$,
6. Obtain the standardized normal variate, say z_0, corresponding to the truncation point x_0 by putting $z_0=(x_0-\hat{\mu}_x)/\sqrt{\hat{V}_x}$.
7. From tables of the normal distribution, find $p_0=\Pr(Z\leq z_0)$, the area under the tail of a standard normal curve to the left of z_0.

8. Hence obtain

$$\hat{s}^* = \frac{s}{1-p_0} \tag{3.19}$$

the estimated number of species in the community including those absent from the sample.

9. Compile Table 3.5 as follows; the Roman numerals refer to the columns in the table which show:

(i) The r values or groups of r values.

(ii) $\log_{10}$ (upper boundary of each group); for example the entry for $r=3$ is $\log_{10} 3.5 = 0.54407$.

(iii) The variate values in (ii) in standardized form. Thus (iii) = $[(\text{ii}) - \hat{\mu}_x]/\sqrt{\hat{V}_x}$.

Table 3.5 The fitting of a truncated lognormal distribution to the data in Table 3.4. Estimation of the needed parameters is described in the text, which also gives the column headings.

$\bar{x} = 1.14276$; $\sigma^2 = 0.44238$; $\gamma = 0.21222$; $\hat{\theta} = 0.02805$;

$\hat{\mu}_x = 1.102266$; $\hat{V}_x = 0.500853$; $\sqrt{\hat{V}_x} = 0.707709$

$z_0 = -1.98287$; $p_0 = 0.02369$; $\hat{s}^* = 115.742$

r (i)	(ii)	(iii)	(iv)	Frequencies Expected (v)	Observed (vi)
0	0.30103	−1.98287	(2.742)	(2.742)	—
1	0.17609	−1.30869	9.686	6.94	16
2	0.39794	−0.99522	18.497	8.81	10
3	0.54407	−0.78874	24.900	6.40	5
4	0.65321	−0.63452	30.425	5.53	10
5–7	0.87506	−0.32104	43.298	12.87	10
8–11	1.06070	−0.05874	55.160	11.86	20
12–17	1.24304	0.19891	66.995	11.84	6
18–23	1.37107	0.37982	74.996	8.00	10
24–35	1.55023	0.63298	85.259	10.26	7
36–72	1.86034	1.07116	99.301	14.04	6
72–144	2.15987	1.49440	107.925	8.62	4
>144	∞	∞	115.742	7.83	9
				113.00	113

$\chi^2 = 33.8$ $P(\chi^2 \mid 9) \simeq 0.0001$

(iv) $\hat{s}^* \times$[Normal probability integral for the standardized variate in (iii)]. Thus the column gives cumulated expected frequencies.

(v) Differences between successive entries in (iv). Hence these are the desired expected frequencies of the r values in (i).

(vi) The observed frequencies.

10. Judge the goodness of fit of the expected to the observed frequencies by a χ^2 test. The number of degrees of freedom is three less than the number of frequencies compared since two parameters have been estimated from the data. As may be seen from Table 3.5, the fit in this case was poor.

One may also obtain estimates of the sampling variances of $\hat{\mu}_x$ and $\sqrt{\hat{V}_x}$. From Cohen's (1961) Table 3 (left half: for truncated samples) obtain μ_{11} and μ_{22} corresponding to η $(= z_0)$. Then

$$\widehat{\text{var}}\,(\hat{\mu}_x) = \frac{\mu_{11}\hat{V}_x}{s} \qquad \text{and} \qquad \widehat{\text{var}}\,(\widehat{\sqrt{V_x}}) = \frac{\mu_{22}\hat{V}_x}{s}.$$

For the data in our Table 3.5 in which $z_0 \simeq -1.98$, it is found that

$$\mu_{11} = 1.21486 \qquad \text{and} \qquad \mu_{22} = 0.751658.$$

Hence

$$\widehat{\text{var}}\,(\hat{\mu}_x) = 0.00538 \qquad \text{and} \qquad \widehat{\text{var}}\,(\widehat{\sqrt{V_x}}) = 0.00333.$$

The lognormal distribution, like the normal, has two parameters. These may be expressed as its own mean and variance, say μ_y and V_y, or as the mean and variance (μ_z and V_z) of the normal variate $z = \ln y$. Equations (3.17) show how these moments are interrelated. The mode of the lognormal is $\exp[\mu_z - V_z]$. Clearly, the greater the value of V_z for given μ_z, the greater V_y, the longer the tail of the distribution, and the smaller the modal value of y. Figure 3.3 shows two examples.

Although the distribution has only two parameters, there is a third "unknown" attaching to any body of empirical data, namely the location of the truncation point relative to the mode or, equivalently, the value of s^*, the number of species in the whole community. Clearly, if only a small sample is taken, s will be low, $s^* - s$ will be high and the species for which $r = 1$ (or for which "continuous r" lies in the range $(\frac{1}{2}, 1\frac{1}{2}]$) will be ones that are comparatively abundant. The truncation point may even fall to the right of the mode implying that those species whose abundance is most typical of the community are too sparse (relative to sample size) to be represented. Now suppose that sample size is increased manyfold: s will be larger, $s^* - s$ will be smaller, and even quite rare species will probably be represented by $r = 1$ or more members. Increasing sample size thus has the effect of shifting the truncation point to the left. As will be seen from steps

6, 7, and 8 in the recipe above, the location of the truncation point determines the area under the fitted curve corresponding to the abundances of *un*represented species, and this area is proportional to $s^* - s$.

In fitting a theoretical lognormal to empirical data we therefore have to estimate the distribution's parameters, μ_x and V_x as in step 5; and we must also estimate s^* as in step 8. However, though we can obtain estimates of var $(\hat{\mu}_x)$ and var $(\widehat{\sqrt{V_x}})$, an estimator of var $(\hat{s}^*)$ has yet to be derived. The present lack of such an estimator is unfortunate as s^* is obviously the most interesting of the three numbers to be estimated.

The problem of estimating s^* from a sample is, indeed, far from straightforward. It can be done iff (if and only if) the observed species-abundance distribution is well fitted by the truncated negative binomial distribution or the lognormal distribution. In the former case (3.5) gives an estimator of s^* and in the latter case, (3.19). But for neither estimator is the sampling variance yet known so that only a point estimate of s^* can be obtained, not an interval estimate; thus judgment as to the precision of an estimate, and comparisons among different communities, are impossible. There is one other possibility. If the observed distribution is of lognormal form and if, instead of fitting an "ordinary" (i.e. continuous) lognormal distribution in the way described above, one fits a Poisson lognormal [with terms as in (3.18)], then an estimate of s^* *and of its variance* can be found (Bulmer, 1974). But an unfortunate fact is, as Bulmer's examples show,

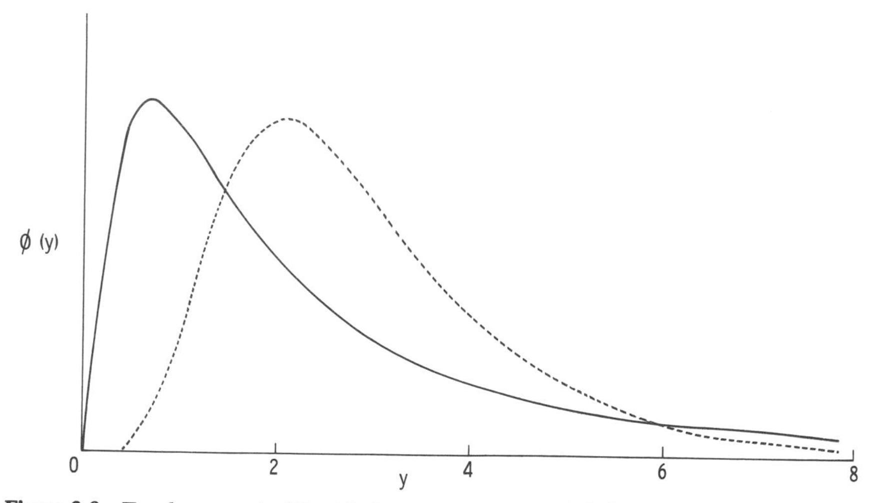

Figure 3.3 Two lognormal pdf's with the same mean, $\mu_y = 3$. Solid line: $\mu_z = 0.5968$; $V_z = 1$; $V_y = 15.46$; mode at $y = 0.67$. Dotted line: $\mu_z = 0.9736$; $V_z = 0.25$; $V_y = 2.56$; mode at $y = 2.06$.

that estimates of s^* obtained by fitting the Poisson lognormal and the continuous lognormal (which Bulmer calls the "grouped lognormal") are discrepant. Thus, as remarked in Section 1.6, estimates of s^* rarely inspire confidence. The whole problem is ripe for further investigation.

Another such problem is that of whether rare species are, in general, fewer in number than somewhat more abundant species so that f_r has a mode for some $r>1$ (as those who fit the lognormal suppose); or whether f_r decreases monotonically with increasing r (as those who fit the logseries suppose). The latter is the older hypothesis and was first advanced formally in the well-known paper of Fisher, Corbet and Williams (1943). The suggestion that species abundances might be lognormal is due to Preston (1948, 1962) who provides a wealth of interesting examples and inferences stemming from them. Other examples are given by Bliss (1965). Preston (1962) also noticed an apparent relationship between the summed abundances of all the species in a community (N) and the total number of species (s^*).

This relationship has been thought to have important theoretical implications for ecology and has given rise to hypotheses concerning the so-called "canonical" distribution of species abundances. However, it has been shown by May (1975) that the apparent relationship stems, in all likelihood, merely from the mathematical properties of the lognormal distribution and is without biological significance.

Chapter 4

Testing Hypotheses About Species Abundances

4.1 Introduction

In Chapters 2 and 3 we discussed some of the models that might explain observed species-abundance distributions. However, inventing a model that might explain a body of data, and deciding whether one is justified in concluding that it does, are two distinct operations. In this chapter we consider the latter.

In order to choose or devise an appropriate way of testing a model it is helpful, at the outset, to answer three questions. These are:

1. Is the model under test a resource-apportioning model (case A) or a statistical model (case a)? For the purpose of the present discussion we treat the niche preemption and broken stick models (cf Sections 2.2 and 2.3) as resource-apportioning, but not the overlapping niches model (cf Section 2.4) which will be regarded as a statistical model. The reason for doing this is that only the first

two models assume that species abundances are mutually dependent. The overlapping niches model resembles the statistical models in assuming them independent.

2. Do the available data yield an observed ranked-abundance list (case B) or an observed species-abundance distribution (case b)? The distinction is discussed in Section 2.1. Usually a fully censused small community, or a small sample from a large community will give data of type B; whereas a large sample from a large community will give data of type b.

3. Is s^*, the number of species in the whole community known (case C), or unknown (case c)? Fully censused communities obviously yield data of type C. So, often, do samples from small or thoroughly studied communities; even when $s < s^*$ (where s is the number of species in the sample), s^* may be independently known. Samples from large communities usually give data of type c.

Each of the three questions above can be answered in two ways. Therefore there are eight distinct sets of circumstances to consider:

Case ABC. This case is discussed in Section 4.2.

Case aBC. This case is discussed in Section 4.3.

Cases abC and abc. In these cases the model is statistical and the data sufficient to permit compilation of an observed species-abundance distribution. One can therefore test the fit of a theoretical distribution to the observations by doing an ordinary χ^2 goodness-of-fit test; s^* may or may not be one of the parameters estimated from the data. These cases are exemplified by the tests carried out in Chapter 3, where theoretical distributions (truncated negative binomial, gamma, logseries, and lognormal) were fitted to large bodies of data. They will not be considered further here.

Cases ABc and aBc. If s^* must be estimated from the data and the data themselves consist of no more than a ranked-abundance list, there is no satisfactory way of judging the fit of a model to observations.

Cases AbC and Abc. These cases are so unlikely to arise in practice that they do not warrant consideration. Thus it is unreasonable to entertain a resource-apportioning model in the context of a "large" community; if the community is large in the spatial sense the habitat is likely to be heterogeneous at least to some degree, in which case the relative abundances of the species must be affected by the relative areas (or volumes) of different kinds of habitat; and if the community is large in the sense that the taxocene comprising it is large, and contains markedly dissimilar species, they are unlikely to be sharing a single limiting resource.

To conclude this chapter, a new resource-apportioning model will be described in Section 4.4. The model bears the same relation to the lognormal distribution that MacArthur's broken stick model bears to the exponential distribution (cf Section 2.3). The reason for discussing it in this chapter is that it demonstrates convincingly the difficulty of arguing from numerical observations to ecological explanations.

4.2 Testing Resource-Apportioning Models

Consider a ranked-abundance list from some community under investigation. It makes no difference to the arguments below whether the list summarizes data obtained from a complete census or from a sample.

Notice first that to test the fit of a hypothetical model, it is *not* permissible to fit a theoretical frequency distribution to the "frequencies" of the different species as listed. These "frequencies" are not frequencies in the usual sense; they are measures of the "sizes" or abundances of the species encountered, in other words, observations on the variate whose distribution is being investigated. If the model under test assumes the species to be independent of one another, and if the observed abundances are all different, then each observed variate value has an observed frequency of unity. (This is a paraphrasis of comments in Section 2.2; it is worth repeating because misunderstandings on the subject are so common).

However, if one tests a resource-apportioning model, that is, if one is hypothesizing that the abundances of the species result from the way in which some common limiting resource has been shared among them, then one is *ipso facto* assuming that the abundances are *not* mutually independent. In this case the list of s abundance values should be regarded, not as s different observations of a scalar variate but as *one observation of an* s-*element vector-valued variate*. It can be represented by one point in an s-dimensional coordinate frame.

It follows that if the "truth" of a resource-apportioning model is to be judged, several s-species communities must be observed. One will not suffice. Indeed, to test such a model, one needs two things:

1. Observations on several s-species communities each of which yields a ranked-abundance list; equivalently, each gives one observation of the vector-valued variate, or one observed point in s-space;
2. Knowledge of the theoretical s-variate distribution that the model predicts.

Now recall the two resource-apportioning models under consideration, the niche preemption model and the broken stick model (cf Sections 2.2

and 2.3). For the former, the theoretical (multivariate) distribution of ranked-abundance lists has not been derived. We do know, however, that the model is unlikely to be generally applicable, since observed ranked-abundance lists rarely resemble the expected list predicted by the model. The single point in s-space needed to represent such a list can be portrayed in two dimensions by a line: this is the line joining observed points when one plots y_i, the abundance of the ith species, against i (see Figure 2.3*b*, page 29). If y_i is plotted on a log scale, the expectation (according to the niche preemption model) is that this line will be straight. Such empirical data as are so far available (for example, in Whittaker, 1972) suggest that this is seldom the case. Further investigation of the niche preemption model seems unprofitable.

Now consider the broken stick model. In this case the expected s-variate distribution of ranked species abundances is known. Contemplation of it leads to an interesting and (perhaps) surprising conclusion, as we now show. For simplicity it will be assumed that abundances are treated as continuous quantities though this is not necessary to the argument.

Recall that according to the broken stick hypothesis, the abundances of the species in an s-species community are represented by the lengths of the pieces into which a unit stick would be divided by breaks at $s-1$ randomly located points. That is, these points are rectangularly distributed on the interval (0, 1). Now let us write $(y_1, \ldots, y_s)$ for the relative lengths of the pieces or, equivalently, for the relative abundances, in any order, of the s species. $\left[\text{Since } \sum_{i=1}^{s} y_i = 1\right.$, the vector-valued variate could be portrayed by a point in $(s-1)$-space and we could omit the element $\left.y_s.\right]$ Then it can be shown (Wilks, 1962, page 237; and see Eberhardt, 1969) that the pdf of this variate is

$$f(y_1, \ldots, y_s) = (s-1)!$$

inside the $(s-1)$-dimensional unit cube $\{(y_1, \ldots, y_{s-1}): 0 < y_i < 1, i = 1, \ldots, (s-1)\}$ and zero outside this region. We now see that the probability density $f(y_1 \cdots y_s)$ is constant throughout the region in which it exceeds zero; or equivalently, *that all possible ranked-abundance lists are equiprobable.*

This amounts to saying that, according to the broken stick hypothesis, any species-abundance distribution is as likely as any other. Thus the relative abundances of the species in any *one* s-species community, whatever they are, can neither support nor cast doubt upon the hypothesis. A conceptual experiment will clarify this. Suppose a card is drawn at random from a well-shuffled deck; (score 1 for an ace, . . . , 13 for a king).

A card of any denomination is equiprobable; that is any of the possible scores 1, 2, . . . , 13 are equiprobable. The "expected" score is 7 but one would have no reason to expect a "7" rather than any other card on any given occasion. The "expected" value is merely the expected mean in infinitely many repetitions of the drawing.

It should now be clear that a census of the individuals in a single community can never provide evidence either for or against the broken stick model. In asserting that, according to the model, the expected abundance of the ith species in an s-species community is

$$E(y_i) = \frac{1}{s} \sum_{x=i}^{s} \frac{1}{x} \qquad \text{[as in (2.4)]},$$

we are not predicting that this will be the size in a particular community, but only that this is what we expect the average to be over many s-species communities. Also, nothing in the model stipulates that this ith species shall be the same species in every case, even when all s-species communities sampled have identical species complements.

This brings us to a most important point concerning resource-apportioning models. When such a model is proposed, is one asked to assume that the resources available is some one particular community have been divided up among the species in the specified manner; (we shall call this a type *I* model)? Or that, over evolutionary time, the several species have become adapted to tolerance ranges (or resource ranges) whose relative sizes are what the model specifies (a type *II* model)?

For clarity, it is worth expressing this important distinction in two other ways as well. Thus one can say: (1) that a type *I* model predicts the outcome of a local battle among competing species, whereas a type *II* model predicts the outcome of a global battle; or (2) that a type *I* model predicts the sizes of the species' "realized niches" in each community and a type *II* model predicts the sizes of their "fundamental niches." (These concepts will not be mentioned further here; they are discussed in detail in Section 7.2).

Whether a model is to be regarded as type *I* or type *II* is often not made explicit. The niche preemption model is presumably always intended as a type *II* model. As to the broken stick model, it is interesting to reconsider the discussion above (showing that, according to the model, all possible species-abundance distributions are equally common) as it is affected by the distinction we have just made between models of types *I* and *II*.

Regarded as a type *I* model, the broken stick hypothesis is that there will be no necessary resemblance between different communities. The positions of the species in a ranked abundance list, as well as their relative abundances, are wholly arbitrary. However, suppose we were to average the observations from a large number of communities, all comprising the same s species, say $A_1, \ldots, A_s$. Then the two different ways in which this averaging could be done give different results: (1) If species identities were ignored and we averaged the abundances of the ith commonest species in each community (whichever species it was) this average would tend to $E(y_i) = (1/s) \sum_{x=i}^{s} (1/x)$ for $i = 1, \ldots, s$. (2) However, if species identities were taken into consideration, that is, if the relative abundance of species A_i (regardless of its rank) were averaged over all the communities, then its average abundance would tend to $1/s$ for $i = 1, \ldots, s$; hence, on averaging over many communities, all s species would be found to be equally abundant.

Regarded as a type *II* model, the broken stick hypothesis is that the relative proportions of any specified set of s species, when they occur together and form a "community," are predetermined. Different communities (or, as some would prefer, different realizations of the same community-type) with these same species as members will have, apart from sampling differences, identical ranked-abundance lists with the species in the same order. In terms of the playing card analogy, it is as though one card had been chosen and replicates then made of it, one for each "realization" of the community. However, the card picked for replication could have had any denomination; it need not have been a "7." Correspondingly, the relative abundances of the member species of a given species-set may have any form whatever (under the type *II* broken stick hypothesis), but different examples of the same set will resemble one another.

It follows that the type *I* version of the broken stick model is untenable and the type *II* version untestable.

Disregarding the distinction between types *I* and *II*, it might be argued that since the broken stick model predicts a species-abundance distribution of negative exponential form (cf Section 2.3), a test is possible after all. This is true: however, it entails regarding the model as statistical rather than resource-apportioning, that is as belonging either to case *aBC* or to case *abC* as defined in Section 4.1, and acting accordingly.

There is another way of demonstrating the mathematical equivalence of the resource-apportioning form and the statistical form of the broken stick model. In Section 2.3 we showed the equivalence of these two statements: (1) the model predicts an *expected* ranked-abundance list as specified by (2.4) (see page 24); and (2) the model predicts a negative exponential species-abundance distribution. A third statement, equivalent to (1) and

(2) is (as we have argued above) given by: (3) all possible ranked-abundance lists are equiprobable. We now show, heuristically, the equivalence of statements (3) and (2).

Contemplate an exceedingly long stick, and a very large value of s, the number of pieces into which it is to be broken. The $s-1$ randomly located breakage points that (according to the model) must be marked on the stick represent a Poisson point process in one dimension. This statement will be recognized as the same as (3); and as is well known from elementary renewal theory (for example, see Cox, 1962) the lengths of the distances between points are negative exponential variates as stated in (2).

4.3 Testing and Comparing Statistical Models

In this section we consider the fitting of statistical models to small bodies of data; and comparisons among different bodies of data.

The literature on species abundances falls into two distinct classes. Papers of one class are found in biological journals and describe the correspondence (judged intuitively) between the predictions of resource-apportioning models and ranked-abundance lists obtained by observing small communities. [The papers by King (1964) and Whittaker (1972) are examples.] As we have seen in Section 4.2, objective comparisons are difficult, and in the case of the broken stick model impossible.

Papers of the other class—they often appear in the statistical literature—discuss the species abundance distributions predicted by statistical models, and test their fit to data obtained from very large collections, often too large to be regarded as natural units. [The papers by Bliss (1965) and Bulmer (1974) are examples.]

Little or no work has been done on the applicability of statistical models to data from small, and it is hoped natural, communities. It is worth considering whether this would be illuminating. To argue that statistical models are less realistic than resource-apportioning ones seems unjustified. The latter can only be described as realistic in the sense that they are expressed in terms of concrete, easily visualizable objects such as sticks. It would require an impossible act of faith to imagine any true correspondence between the mechanism of such a model and actual ecological processes. Thus, if models are to be fitted at all, there is no good reason to look upon statistical ones with disfavor.

Consider, therefore, the fitting of statistical models to ranked-abundance lists obtained from restrictively defined communities with comparatively few species, say between five and thirty. When a statistical model is fitted and no thought of resource apportionment arises, the species are treated as independent and a ranked-abundance list is merely an ordered random sample of observed abundances. It is perfectly possible to choose a theoretical distribution, estimate its parameters, plot its cdf and judge its goodness of fit to the sample cdf by means of a Kolmogorov-type test. Methods for doing such tests are described in Conover (1971). When the number of species and hence of sample values is small, any test inevitably lacks power but this is unavoidable. The objects sampled are species, and sample size can be increased only by taking account of additional species, not by counting more individuals of the same species; in other words, by redefining the community under study, and perhaps destroying its unity in the process.

Therefore we must now consider what ecological insights might be gained by carrying out the program outlined above in spite of the impossibility of performing powerful tests. If our aim is to search for hitherto undetected uniformities in nature this might, perhaps, be a way of doing it. Explaining such uniformities (if they exist) can wait until they have been found. But the great risk of detecting spurious uniformities must be guarded against. Thus if we were to find, for instance, that truncated lognormal distributions with various parameter values fitted the data in the majority of cases, this would probably tell us less about natural communities than about the multitude of shapes that the family of truncated lognormals can assume. The search for a family of pdf's that will fit the empirical species-abundance distributions of an array of dissimilar and unrelated communities does not seem worthwhile.

A more profitable approach (it seems to me) is to compare fairly similar communities with one another in order to determine whether their species-abundance distributions tend to be of the same form even when the communities are far from homogeneous in species composition. An example will illustrate. Wiegert (1974) has presented data on the microarthropod fauna of plant litter in old-field communities with different vegetation cover. He collected specimens over a period of one year by leaving fiberglass mesh bags filled with plant litter of various kinds in the different areas whose communities were to be compared. We shall here treat, as two comparable communities, the totals of mites (Acari) collected from two of the fields, one with vegetation dominated by broomsedge (*Andropogon* spp.) and the other with *Lespedeza cuneata* as dominant. Table 4.1 and Figure 4.1 summarize the observations. The table shows (above) a comparison of the abundances of the mite species in the two communities; a homogeneity test shows them to be highly inhomogeneous. In the lower part of the table are given the species-abundance distributions for the two communities; they have been used to plot the observed cdf's shown in Figure 4.1 which are obviously very similar.

Table 4.1 The homogeneity test [a]

Mite species	Numbers collected in fields of: Broomsedge	*Lespedeza*	
Asca piloja	30	79	
Ambyseus sp	49	54	
Eupodes sp	22	9	
Lorryia sp	8	3	
Passalozetes sp	26	1	
Tectocepheus sp	0	12	$\chi^2 = 210.4$
Scapheremaeus sp	5	12	$\Pr(\chi^2 \mid 11) < 10^{-6}$
Galumna spp	3	31	
Peloribates sp	166	26	
Trhypochthonius sp	49	88	
Cultroribula sp	15	5	
Miscellaneous uncommon spp	45	61	

Species-abundance distributions Each list shows values of r, the number of individuals in each species. If $f_r > 1$, its value is shown in parentheses after r

Broomsedge field (27 spp): 1(7), 2(3), 3(3), 4(3), 5, 6, 8(2), 15, 22, 26, 30, 49(2), 166
Lespedeza field (29 spp): 1(6), 2(6), 3(2), 4(3), 5(2), 9, 11, 12(2), 13, 26, 31, 54, 79, 88

[a] Data from Table 5 of Wiegert (1974) on the mite communities in plant litter in two fields, one dominated by broomsedge and the other by *Lespedeza*.

Probably the best test to judge whether the distributions can be regarded as the same is the Smirnov test. The two cdf's are plotted and the test statistic is the greatest vertical distance between them. The observed value of the statistic (which is 0.09 in Figure 4.1) is then compared with the quantiles of its expected distribution (tables are given in Conover, 1971) to judge whether an improbably large value has been observed. In the present case the critical value of the statistic is 0.286 for a test of size $\alpha = 0.2$ and we appear to be justified in concluding that the two communities have identical species-abundance distributions.

This conclusion, and the test leading to it, need critical appraisal. Consider the test first. Three difficulties arise in doing such a test:

1. As always when data on species-abundance distributions are gathered, sampling intensity has a pronounced effect on the form of the

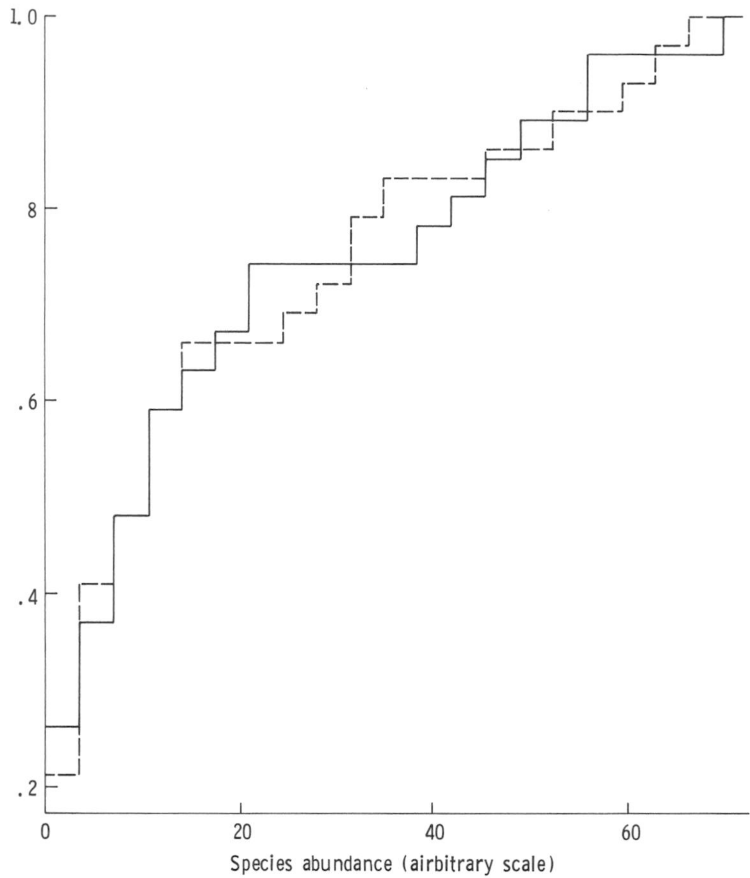

Figure 4.1 The cdf's of two observed species-abundance distributions. Data from Wiegert (1974). The abscissa shows the ranks (in the combined data from both communities) of the species abundances; the absolute values are immaterial.

observed distribution. Clearly, two samples of different sizes, even though they be drawn from the *same* population, can have markedly different species-abundance distributions. The difference will be especially conspicuous if the larger sample contains, and the smaller lacks, those species whose abundances have the modal value for the whole community; for then the larger sample but not the smaller will yield a "humped" distribution. Unless sampling procedures are standardized in some way, therefore, there is a risk that empirical species-abundance distributions will be adjudged different when in fact they are the same.

2. When species-abundances are discrete variates (i.e., counts of individuals per species rather than measurements of biomass) the Smirnov

test is conservative (Conover, 1971). That is, the size of the test (equivalently, the risk, α, of making a type I error) is somewhat smaller than it would be if the variate were continuous. This has the effect of reducing the power of the test and increasing the risk of concluding that the cdf's being compared are the same when they are not (a type II error). This source of error is probably negligible compared with (1).

3. No test has so far been developed permitting comparisons among more than two cdf's when sample sizes are unequal. Sample size in this context is, of course, the number of species in each sample, and hence is something that cannot be decided by the investigator.

So much for the statistical difficulties of comparing empirical species-abundance distributions; there are also difficulties of interpretation. In particular, is the null hypothesis reasonable? When the cdf's of two observed species-abundance distributions are being compared, as in Table 4.1, the null hypothesis is that the relative proportions of the species comprising the communities are the same even though the species themselves are different. Thus in the example, only 19 mite species were common to both communities and an additional 18 species (mostly with few individuals) occurred in only one (as opposed to both) of the communities. Also, as the test in the table shows, the communities did not match in the relative proportions of their shared species. The hypothesis* that the species-abundance distributions of the two communities may be the same notwithstanding these pronounced differences between them exemplifies one of the chief tenets of many students of ecological diversity. This is the hypothesis that communities have a "structure" that is to some extent independent of what their component species are. Whether they have or not is a problem still to be solved. The matter is discussed more fully in Chapter 7. It suffices to say here that an affirmative answer would obviously be far more convincing if it were based on evidence obtained by comparing a very large number of pairs of fairly similar communities than on evidence from only a few pairs of markedly dissimilar communities. Before more can be said, much needs to be done both in data gathering and in refining the statistical tests.

4.4 The Sequential Breakage Model

In Section 2.3 it was shown that the species-abundance distribution predicted by MacArthur's broken stick model of resource apportionment is the negative exponential distribution. In Section 3.3 it was shown that the

* Wiegert's (1974) paper is not concerned with this hypothesis. His data have been used here merely for illustration.

species-abundance distribution predicted by a somewhat different stick-breaking process is the lognormal distribution. It is therefore interesting to consider the ranked-abundance list predicted by the latter process.

Since two broken stick models are to be contrasted, they should be distinctively labeled. In the MacArthur model the several breaks in the stick are envisaged as being made simultaneously whereas in the model leading to the lognormal distribution they are made sequentially. Therefore we shall speak of the *simultaneous breakage model* and the *sequential breakage model.* For the simultaneous breakage model, we derived the expected ranked-abundance list first and the species-abundance distribution second. For the sequential breakage model we have already considered the species-abundance distribution (which is lognormal) and we now examine its ranked-abundance list.

Consider the simplest sequential breakage model; (it is one of a whole family of models all of which yield the lognormal as the limiting distribution of the lengths of the parts into which a stick is broken; cf Section 3.3). Recall that the unit stick was broken once, at a random point; one of the two parts was picked at random and itself broken, again at a random point, so that there were three parts; and so on. Let us obtain the expected lengths of the parts for low values of s, the number of parts (or of species). After one break, which gives $s=2$, we find from (2.4) that

$$E(y_1)=0.75 \qquad \text{and} \qquad E(y_2)=0.25,$$

where y_i is the size of the ith largest part (or abundance of the ith most abundant species).

Now suppose a break yielding parts with these *expected* lengths has been made. The expected result of the second break (after which there will be three parts) depends on whether the larger or smaller part is broken next. The conditional expectations of the lengths of the parts (in order of rank) are as follows. If, after the first break the larger part is chosen for the second break, then

$$E(y_1 \mid y_1)=0.75^2=0.5625; \qquad E(y_2 \mid y_1)=0.25;$$
$$E(y_3 \mid y_1)=0.75\times 0.25=0.1875;$$

and if the smaller part is chosen, then

$$E(y_1 \mid y_2)=0.75; \quad E(y_2 \mid y_2)=0.25\times 0.75=0.1875;$$
$$E(y_3 \mid y_2)=0.25^2=0.0625.$$

Since each of these outcomes is equally likely, the unconditional expected ranked-abundance list when $s=3$ is the mean of these two conditional lists, namely

$$(0.6563 \quad 0.2187 \quad 0.1250).$$

In the same way, to determine the conditional expectations following the third break, we can calculate $E(y_i | y_j)$ for $i = 1, \ldots, 4$; $j = 1, 2, 3$. Averaging the $3! = 6$ equiprobable lists gives, as the unconditional expected ranked-abundance list when $s = 4$,

$$(0.6016 \quad 0.2005 \quad 0.1354 \quad 0.0625).$$

For $s = 5$ the list is

$$(0.5640 \quad 0.1938 \quad 0.1322 \quad 0.0729 \quad 0.0371).$$

And so on. When s becomes large, the expected abundances are indistinguishable from s independent random variates from a lognormal distribution.

When s is small and data from only one community are available, this model is impossible to test for the same reason that the familiar simultaneous breakage model is untestable. Therefore, field ecologists are unlikely to have any use for it. However, both the broken stick models merit the attention of laboratory investigators who can culture numerous replicate communities of a few chosen species living under controlled conditions. Averaged data from replicates will yield observations that can be compared with the models' predictions in a way that data from communities in the field (which are always unique in at least some respects) cannot be.

Chapter 5

Diversity and Spatial Pattern

5.1 Introduction

Students of diversity search for general insights, and for conclusions that apply to ecological communities of all kinds. Whatever group of organisms one worker specializes in, whether it be trees, insects, corals or slime molds, the hope and intention is that the conclusions can be extrapolated. Therefore it is most important that field data and the communities they come from be comparable.

This raises problems of extraordinary difficulty. A decision as to whether or not two communities can or cannot reasonably be compared is necessarily subjective. In many such cases, opinions (if they were sought) might well be evenly divided. There are three independent sources of difficulty, all equally perplexing, since to define, precisely, the boundaries of the "community" being investigated, one must specify: (1) the taxonomic limits of the groups of organisms (the taxocene) to be treated as community members; (2)

the time interval during which observations are made, and through which the community is thought of as persisting as an entity; and (3) the region of space the community occupies. This chapter will be concerned with matters arising from (3) but it is worth saying a few words about (1) and (2).

First as to (1), the noncomparability of different taxocenes is well known. They may be broadly or narrowly defined: for example, given the contents of a light trap for insects, one might regard every member of the class Insecta caught in it as belonging to a community to be studied, or one might select some subset of ordinal or familial rank; how should one choose? Consistency in the choice of the hierarchical rank of taxocenes to be regarded as forming communities does not ensure comparability, however. There will probably never be a wholly satisfactory answer to such a question as: are the species in an order of angiosperms more or less similar to one another than the species in an order of birds? Therefore one can never state, dogmatically, that two taxonomic orders, say, are or are not comparable. Further, quite apart from taxonomic inclusiveness, can one treat as comparable two communities containing different numbers of trophic levels? Most workers would not regard wood-warblers-plus-insects in a deciduous woodland in spring as forming a single entity, although they would so regard the total of all the fishes in a lake. But if the fish community is to include prey as well as predators, why not the warbler community?

Further difficulties arise over taxonomic uncertainties and the perennially irreconcilable disagreements of taxonomic splitters and lumpers; over the treatment to be accorded to polymorphic species and species exhibiting sexual dimorphism; over "difficult" genera such as (among North American plants) *Rubus* and *Vaccinium*; over plants that form hybrid swarms; over organisms that occur as asexually reproducing clones such as aphids, and the majority of plants; and over species that, in a community whose area contains a variety of habitats, occur as a number of ecologically distinct populations or ecodemes. All these problems can be mentioned but not solved. Most of them are aspects of a more general problem: since community structure is bound up with the concept of the ecological niche, what entity should be treated as the "proprietor" of a niche? The latter question will be taken up in Chapter 7.

Point (2) concerns the temporal extent of a "community." A census of the trees in a few hectares of forest can be made in a few days and gives a snapshot of what may be, in terms of the trees' life spans, a very unstable community; but the speed, relative to these life spans, with which the census can be carried out ensures that no appreciable change in community structure will occur during the observation period. In contrast, the insects caught in a whole night's run of a light trap is likely to include many species that are never simultaneously active; and an assemblage of fossils treated as a "fossil community" may contain individuals whose lifetimes were separated by centuries. Whether such time-smoothing is an aid or a hindrance to the discovery of how a community's component species coexist and interact is debatable. Smoothing may mask chance, short-period anomalies in community composition that could render a snapshot census untypical; but it may obscure important cyclical variations or long-term successional trends. And it may lead to an overestimate of the numbers of species present and active at any one instant of time.

We come now to problem (3). A student of community structure must necessarily specify the spatial boundaries of a community chosen for study. Sometimes a community creates itself, within predetermined boundaries, as the result of an experiment. Examples are: the diatoms settling on a prepared surface (cf Section 3.3); the mites colonizing a litter bag (cf Section 4.3). For a "ready-made" community, the boundaries may coincide with those of an *island*; the word, in its ecological sense, connotes any habitat entirely surrounded by another with strikingly different properties, the two being separated by fairly abrupt boundary. Or else the boundaries may be deliberately drawn, to mark off a representative part of some large, uniform community. Use of words such as *uniform* and *representative* to describe a chosen area is apt to trigger futile and inconclusive debate between the critics and defenders of a particular research project. The inescapable fact is that until a piece of research is far advanced one often does not know whether an area is uniform, and cannot judge how small a piece of it is representative (even if universally acceptable definitions of these terms could be found). Clearly, there is an implicit contradiction in insisting that an area chosen for study be uniform in some sense *before* an investigation of its properties begins.

Therefore, imprecise descriptions, employing loosely-defined terms, are inevitable at the start of an investigation which is when the community to be studied is chosen and its boundaries decided. This, by definition, is the moment at which intuition gives way to exact investigation. Spatial heterogeneity at a hierarchy of scales is present in (probably) all communities and this heterogeneity is itself a form of diversity. It simultaneously affects, and is affected by, interactions within and between species populations. No one standard method for studying spatial diversity exists or could exist; some possible approaches are considered in the succeeding sections of this chapter.

5.2 The Patterns of Mosaics

Most communities of sessile organisms can be thought of as mosaics. The mosaic structure is sometimes clear, as it often is in the vegetation of mountain meadows, bogs, moors, intertidal seaweed beds and some coral communities. In other cases (e.g., forests, mixed grassland) the mosaic patches may be so small, being of the same order of size as the individual organisms, that they can scarcely be described as mosaics in the colloquial sense; but it is still useful to treat them as mosaics. Many animal communities, also, have mosaic patterns in virtue of the patterns formed by the substrates on which they occur. Thus when a beach is a mosaic of sand and mud patches, the snails and clams of the interstitial mollusk fauna sort themselves into separate microcommunities whose pattern matches that of the substrate. Often the substrate is itself a community, of plants. Then an animal community may have a mosaic pattern matching that of the plants, or plant parts, that are the animal species' preferred habitats. An excellent example of such a mosaic (in three dimensions) is given in MacArthur's (1958) classic study of the patterns of five species of wood warblers (*Dendroica* spp) in Maine spruce forests. Each species tended to spend most of the time in its own region within a tree; for instance, the Blackburnian warbler (*D. fusca*) does most of its foraging at the top of a spruce tree on or near the outer surface of the crown. The five warblers had partitioned the forest, considered as a volume wholly occupied by tree crowns and the spaces beneath them, into an intricate mosaic consisting of innumerable replicates (the individual trees) of a single basic pattern.

Now consider how the mosaic pattern of a community affects what we speak of as the community's diversity (in the intuitive sense). Recall that, when spatial pattern is disregarded, a community's diversity is (again intuitively) equivalent to the "uncertainty" attaching to the specific identity of an individual picked at random from it. Thus if H' (or, for a "small" community, H) is used as an index of diversity, it is clear from the index's properties (listed in Section 1.2) that the higher the value of H' the lower the probability that the species of an individual picked at random will be correctly predicted; in other words, the greater the uncertainty surrounding the prediction. But if we take the community's spatial pattern into account, and attempt to predict the species of an individual near (in space) to one that has already been identified, the chance of successful prediction becomes much greater: there is much less uncertainty about the identity of a nearby neighbor than a distant one. Further, for a given distance between the known and unknown individuals, the coarser the pattern of the mosaic the greater the chance of successful prediction.

Let us formalize the foregoing discussion and use it as a basis for devising a measure of "spatial diversity." For clarity, we shall envisage a vegetation mosaic: it is an area wholly occupied by nonoverlapping patches of different phases. The *phases* are the different kinds of vegetation present; a phase may consist of a single species, or a recurring group of species that forms a recognizably distinct microcommunity-type; in a stylized map, each phase would be distinctively colored. The phases are fragmented into *patches*. In a mosaic whose phases are single species (as opposed to microcommunities) a patch of a given phase is a "patch" in the colloquial sense if the species reproduces vegetatively to form extended clones in which distinct individuals are not recognizable; or a "clump" if the species occurs as separate individuals. In either case we shall, throughout this section, treat each patch as a continuous finite area, not as a swarm of points.

Now consider a mosaic made up of s different phases. To define a measure of its spatial diversity we proceed as follows. Let the mosaic be sampled at a sequence of $N+1$ equidistant points along a line transect laid across it, and let the phase in which each point falls be recorded. For example, letting $s=4$ and labeling the phases A, B, C, and D, the observations consist of a list such as

$$AAA\ BB\ C\ B\ DDDDD\ CC\ BBB\ D$$

as shown in Figure 5.1.

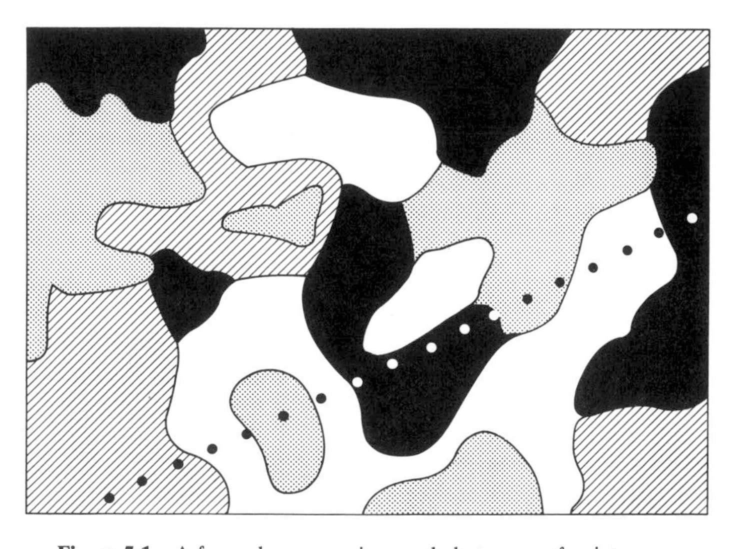

Figure 5.1 A four-phase-mosaic sampled at a row of points.

From the list we can construct an $s \times s$ matrix of *digram frequencies* (Shannon and Weaver, 1949), say $\mathbf{N} = \{n_{ij}\}$ where n_{ij} is the observed frequency with which a point in the ith phase is succeeded by a point in the jth phase. (Examples of digram matrices appear in Tables 5.1 and 5.2.) The total number of these *one-step transistors is* $\sum_i \sum_j n_{ij} = N$; and $n_{ij}/N_i = p_{ij}$ may be taken as an estimate of the probability that a point in the ith phase will be succeeded by a point in the jth phase. The row totals of $\mathbf{N}$, namely $\sum_i n_{ij} = N_i$ (for $i = 1, \ldots, s$) give the total number of sampling points (excluding the last) that fall in phase i; thus $N_i/N = P_i$ is an estimate of the areal proportion of phase i in the mosaic.* Then

$$H' = -\sum P_i \log P_i = -\frac{1}{N}\{\sum N_i \log N_i - N \log N\} \tag{5.1}$$

gives the mosaic's diversity as measured in the usual way, taking into account only the relative areal proportions of the phases. To measure what may be called the mosaic's spatial diversity we can use $H'_{(1)}$ defined as

$$\begin{aligned} H'_{(1)} &= -\sum_i \sum_j P_i p_{ij} \log p_{ij} \\ &= -\frac{1}{N}\left\{\sum_i \sum_j n_{ij} \log n_{ij} - \sum_i N_i \log N_i\right\}. \end{aligned} \tag{5.2}$$

If the observed sequence of phases forms a realization of a simple Markov chain, $H'_{(1)}$ is—to use the terminology of information theory—a measure of the average gain of information as the chain advances one step, or the entropy of the chain (Khinchin, 1957). Regardless of whether the chain is Markovian or not, however, $H'_{(1)}$ as defined in (5.2) appears to be the most convenient expression to use as a measure of spatial diversity. Observe that

$$\begin{aligned} H'_{(1)} + H' &= -\frac{1}{N}\left\{\sum_i \sum_j n_{ij} \log n_{ij} - N \log N\right\} \\ &= -\sum_i \sum_j P_i p_{ij} \log P_i p_{ij}. \end{aligned}$$

This sum is, in effect, the "diversity" of the transitions in the sequence when these transitions are regarded as N "individuals" of s^2 different "species".

* In this chapter we drop the distinction noted in Chapter 1 between a finite and an "infinite" community, and write P_i (or p_{ij}) and N_i/N (or n_{ij}/N_i) interchangeably, as convenient. Thus sampling error, as it affects observations on a single transect, is disregarded. The justification for this will appear subsequently.

Observe also that if the phase encountered at any point on a transect is independent of the phase at the preceding point, that is if $p_{ij}=P_j$ (or equivalently, $n_{ij}=N_iN_j/N$), then $H'_{(1)}=H'$ and this is the maximum value that $H'_{(1)}$ can attain. This accords with intuition: if knowledge of the phase at a given point in the transect does nothing to diminish our uncertainty as to the next point's phase, then $H'_{(1)}$ is as great as H'. But if the phase at a point is not independent of the phase at the preceding point, that is if $p_{ij}\neq P_j$, then our uncertainty about the phase at the next point must be less than our uncertainty about the phase at a point located anywhere, at random, in the mosaic, and consequently $H'_{(1)}<H'$.

This is demonstrated by artificial numerical examples in Table 5.1. Three 4×4 digram matrices are shown, for all of which H' is the same; (natural logs are used in the calculations). In $\mathbf{N}_1$, $n_{ij}=N_iN_j/N$ for $i, j=1, \ldots, 4$ and hence $H'_{(1)}=H'$. $\mathbf{N}_2$ is the matrix that would result if the four species occurred as four large patches (except that one of the A's has been put at the end of the sequence to make the row and column totals identical); $H'_{(1)}$

Table 5.1 Examples of matrices of digram frequencies

$\mathbf{N}_1=$ (rows: First species of a transition; columns: Second species of a transition)

	A	B	C	D
A	16	12	8	4
B	12	9	6	3
C	8	6	4	2
D	4	3	2	1

$H'_{(1)}=1.2798$ nats

$\mathbf{N}_2=$

	A	B	C	D
A	39	1	0	0
B	0	29	1	0
C	0	0	19	1
D	1	0	0	9

$H'_{(1)}=0.1628$ nats

$\mathbf{N}_3=$

	A	B	C	D
A	0	22	11	7
B	20	0	7	3
C	13	7	0	0
D	7	1	2	0

$H'_{(1)}=0.8573$ nats

In all three matrices,

$\{N_i\}=(40\ 30\ 20\ 10)$; $N=100$ and $H'=1.2798$ nats.

is very low. $\mathbf{N}_3$ is the matrix that results when the A's, . . . , D's are deliberately mingled as thoroughly as possible so that two adjacent transect points are never in the same phase; such "supermingling" of the phases leads to less, not more, uncertainty about the phase at a point next to a known point and hence to less, not greater, spatial diversity as shown by the fact that $H'_{(1)} < H'$.

As the foregoing discussion shows, a measure of spatial diversity is easy to devise. We now consider how and when measurements of spatial diversity are likely to be ecologically revealing. The spatial pattern of a many-species community of sessile organisms must obviously have a pronounced effect on within- and between-species interactions and is therefore one of the characteristics that ought to be measured when several such communities are to be compared. Obviously, however, since any one transect yields only a single value of $H'_{(1)}$, without any estimate of its sampling variance, comparisons must be based on data from several transects across each of the mosaics being compared. Then data will be available for nonparametric testing. For example, the Mann-Whitney test (see, for instance, Conover, 1971) can be used to judge whether the spatial diversities of two communities differ at a chosen significance level.

Before calculating $H'_{(1)}$ for each of several transect lists from one area it is important to ensure, however, that the separate lists are homogeneous, that is, that the probabilities (as estimated by the observed relative frequencies) of the different kinds of transitions do not vary among the transects. This defines a *stationary* pattern. All field workers are familiar with the following difficulty that often arises when a study area is being selected and marked out: habitats often turn out to be less homogeneous, on close inspection, than had at first been thought, with the result that attempts to enlarge an area (to mask its internal nonuniformities and make it more representative of a putative larger whole) are thwarted by the fact that such enlargement will bring even worse nonuniformities within the area's boundaries. If a region is crossed by several unrelated environmental gradients with axes in different directions, along which different habitat factors exhibit trends, no internally homogeneous "community" can be said to exist and measurements of spatial diversity are valueless.

To test whether the vegetational mosaic in a study area is internally homogeneous or stationary, therefore, we must obtain transect lists from several randomly located and randomly oriented transects and test them for homogeneity. The test is done as follows. Suppose t transects have been observed, each at $N+1$ equidistant points. There are t different $s \times s$ matrices to compare. Together they form an $s \times s \times t$ contingency table and we wish to test the independence of the t layers.

Let n_{ijk} be the frequency in the (i, j, k) th cell $(i, j=1, \ldots, s;\ k=1, \ldots, t)$. Put

$$\sum_k n_{ijk}=n_{ij\cdot} \qquad \sum_i\sum_j n_{ijk}=n_{\cdot\cdot k} \qquad \sum_i\sum_j\sum_k n_{ijk}=N.$$

To do a likelihood ratio test (see, for example, Wilks, 1962) we require the test statistic $-\ln\lambda$ defined as

$$-\ln\lambda=\sum_i\sum_j\sum_k n_{ijk}\ln n_{ijk}-\sum_i\sum_j n_{ij\cdot}\ln n_{ij\cdot}-\sum_k n_{\cdot\cdot k}\ln n_{\cdot\cdot k}+N\ln N \qquad (5.3)$$

Then under the null hypothesis, $-2\ln\lambda$ is asymptotically distributed as a χ^2 variate with $(s^2-1)(t-1)$ degrees of freedom. It will be observed that (5.3), the formula for a log likelihood ratio, is of the same form as the formulae for "diversity," "information," or "entropy." The test statistic $-2\ln\lambda$ has been called the minimum discrimination information statistic by Kullback, Kupperman, and Ku (1962) who discuss the connexion between information theory and likelihood ratio tests.

(It is instructive to demonstrate, heuristically, why $-2\ln\lambda$ tends to a χ^2 variate. Observe that under the null hypothesis the expected value of n_{ijk} is $\mathscr{E}(n_{ijk})=n_{ij\cdot}n_{\cdot\cdot k}/N$.
Therefore

$$-\ln\lambda=\sum_i\sum_j\sum_k n_{ijk}\ln\frac{n_{ijk}}{(n_{ij\cdot}n_{\cdot\cdot k}/N)}\equiv\sum O\ln\left(\frac{O}{E}\right)$$

where O and E are the observed and expected frequencies in one of the s^2t cells of the $s\times s\times t$ contingency table, and the summation is over all cells. Observe that (provided O and E are numerically close which they will be if the null hypothesis holds), then when

$$O>E, \qquad \ln\left(\frac{O}{E}\right)=\ln\left(1+\frac{O-E}{E}\right)\simeq\frac{O-E}{E};$$

and when

$$O<E, \qquad \ln\left(\frac{O}{E}\right)=-\ln\left(\frac{E}{O}\right)=-\ln\left(1+\frac{E-O}{O}\right)\simeq\frac{-(E-O)}{O}.$$

Then since, under the null hypothesis, $\Pr(O<E)=\Pr(O>E)=0.5$, it is seen that the expectation of $\ln(O/E)$ is

$$\mathscr{E}\left[\ln\frac{O}{E}\right]\simeq\frac{1}{2}\left\{\frac{O-E}{E}-\frac{E-O}{O}\right\}=\frac{1}{2}\,\frac{O^2-E^2}{OE};$$

Therefore

$$\mathscr{E}[-2\ln\lambda]=\mathscr{E}\left[\sum 2O\ln\left(\frac{O}{E}\right)\right]\simeq\frac{O^2-E^2}{E}=\sum\left(\frac{O^2}{E}\right)-N$$

which will be recognized as being identical to $\sum(O-E)^2/E$, the familiar formula for the test statistic for a χ^2 goodness-of-fit test.)

Table 5.2 shows an application of the test to actual field data. The mosaic sampled was the forest floor vegetation in temperate deciduous woodland in early spring and four phases were recognized: bare ground (B), herbs (H), grasses and sedges (G), and tree seedlings (T). Two 5-meter transects were laid down and the phase recorded at 5-centimeter intervals along each of them. As the test in the table shows, the two transect lists appear to have been homogeneous.

Table 5.2 Test of the homogeneity of two matrices of digram frequencies [a]

From Transect 1

	B	H	G	T
B	18	13	1	1
H	10	25	5	3
G	1	3	4	1
T	3	2	0	10

From Transect 2

	B	H	G	T
B	24	13	3	0
H	10	28	3	2
G	4	2	3	0
T	1	1	0	6

$s=4$; $t=2$; $n_{..k}=100$ for $k=1,2$.

$-2\ln\lambda=10.225$

$\Pr\{\chi^2\geq 10.225 \mid \nu=15\}>0.75.$

[a] Data from Pielou, unpublished.

When $H'_{(1)}$ is used as a measure of spatial diversity its value depends, of course, on the distance between the sampling points on the transects. In a stationary mosaic $H'_{(1)}\rightarrow H'$ as this distance is made larger. Choice of the distance to be used must be made on the ground, on a common-sense basis, taking account of the mosaics to be compared. Too short a distance between sampling points makes the work laborious and too long a distance leads to loss of resolving power. Another decision that has to be made concerns the lengths of the transects. For a given number of sampling points, one can use many short transects or a few long ones. Suppose that in an area being sampled there are numerous repeated *pattern units* (the meaning of the term is self-evident; it is discussed further below); then, if the length of the transects is of the same order of magnitude as the mean diameter of the pattern units, the frequencies in the observed digram matrix will not be independent of transect length, for the observed sequence of phases could (in theory, at any rate) be a realization of a

Markov chain of order $n > 1$. To ensure comparable results it probably suffices to sample transects of constant length for all replicates in every mosaic of the set that is to be intercompared.

The need for arbitrary decisions before sampling can begin is of no consequence in practice since values of $H'_{(1)}$ from a single community are of no interest by themselves; they are needed only when two or more communities are being compared. Therefore a procedure for making the observations should be chosen that is appropriate for all the communities under investigation.

Consider now the circumstances in which comparisons of mosaic patterns are likely to be ecologically rewarding. Probably the most interesting mosaics are tessellations of repeated pattern units in which the units depart from strict identity with one another only because of stochastic effects (the omnipresent white noise that bedevils all ecological field observations). Vegetation mosaics of this kind can have two quite different causes: they may occur on *patterned ground*; or they may be endogenously controlled *cyclical mosaics.* Patterned ground (Washburn, 1956; West, 1968) is ground that because of frost action (or occasionally because of dessication) exhibits a microtopographic pattern of conspicuous regularity; it occurs chiefly, but not exclusively, in cold climates. Examples are stone circles, stone polygons, frost-crack polygons, dessication polygons and hummock fields. The patterning is caused by physical (abiotic) factors and imparts to such vegetation as grows on it a corresponding pattern; examples have been described by Watt, Perrin and West (1966) and by Clayton (1966). Cyclical vegetation mosaics, by contrast, can form in habitats of complete uniformity since the patterning is endogenously caused. The phases are species (or microcommunities) that represent different stages in a temporal cyclical succession. The succession is in different stages in different patches which together form a mosaic that can have striking regularity. Examples have been described by, for instance, Watt (1947) in heath vegetation, Harper et al. (1961) in sphagnum bogs, and Anderson (1967) in the vegetation of raised beaches.

Mosaics that are due solely to patterned ground or solely to cyclical succession are easily distinguished, given time, since in the latter the patches are in a state of continuous wandering migration. (These mosaics are stationary, notwithstanding, since every subarea large enough to contain all phases resembles every other such subarea regardless of location.) Interesting, if confusing, mosaic patterns are to be expected where vegetation with a tendency to form cyclical mosaics grows on ground having its own regular microtopographic mosaic; the scales of the two mosaics are not likely to match. Even more confusing patterns are much commoner, indeed ubiquitous; they occur wherever a stationary habitat contains random (as opposed to geometrically regular) microtopographic heterogeneity, and patchy (though not necessarily cyclically patchy) species-populations. This description fits nearly all stationary sessile communities and the chief problem they pose to students of diversity is that of distinguishing abiotic, or exogenous, spatial diversity from endogenous spatial diversity. (One could say that the problem is to distinguish exogenous from endogenous pattern, but the word "pattern" is best avoided here since we are no longer contemplating more or less regular patterns.)

Endogenous spatial diversity is that which would occur even in a perfectly homogeneous habitat; in other words, it is "residual" spatial diversity that would remain if the effects of exogenous spatial diversity could be excluded. Most ecologists are familiar with tiresome debates over whether a calculated value of diversity should be regarded as "within-habitat" or "between-habitat." Such debates are semantic and arise from disagreements as to which rank of entity, in a hierarchy of ranks, deserves to be called a "habitat"; they could be avoided if an objective method existed for recognizing the difference between exogenous and endogenous spatial diversity. Devising such a method will not be easy but is certainly worthwhile.

5.3 Random Mosaics

A mosaic can be described as random if the different patches of which it is formed are randomly mingled. For a given number of phases, given areal proportions of these phases, and a given size scale, a mosaic that is random in this sense has maximum possible spatial diversity; this follows since only in a random mosaic is the phase of any given patch independent of that of all contiguous patches, except for the fact that contiguous patches cannot be of the same phase by definition.

It is therefore worth considering how random mosaics may be recognized and in this section a method will be described for testing the null hypothesis that the sequence of phases observed at equidistant points along a transect have come from a mosaic with randomly mingled patches. It should be pointed out, however, that nothing can be said about the power of the test. It would not be difficult to construct a geometrically regular mosaic that would give the same observational data as a truly random mosaic. Thus the null hypothesis of randomness is not worth testing unless there are prior reasons for entertaining it.

A random mosaic will result whenever the species, or microcommunities, constituting the different phases have patchy spatial patterns that are independently superimposed. The null hypothesis is therefore equivalent to the hypothesis that the phases are independent, in the sense that no two of them are especially prone to occur contiguously; it does not follow, of course, that the organisms in contiguous patches do not interact once chance has brought them into contact.

In the test now to be considered, the relative sizes of the patches are deliberately disregarded. We are interested only in whether they are randomly mingled. Thus suppose a list of the phases sampled along a transect is, for instance,

$$AAA\ B\ AA\ CCCC\ D\ AAA\ BB\ A\ CCCC\ DDD \cdots;$$

then, before beginning the analysis, we "collapse" this sequence to

$$A\ B\ A\ C\ D\ A\ B\ A\ C\ D \cdots.$$

in which every patch encountered, regardless of its size, is represented by one symbol. It is assumed that all s phases in the mosaic have been sampled at least once.

Under the null hypothesis the collapsed sequence is a simple Markov chain with matrix

$$\mathbf{Q}=\{q_{ij}\} \qquad \text{where} \qquad q_{ij}\begin{cases} >0 & \text{when} \quad i\neq j \\ =0 & \text{when} \quad i=j \end{cases}$$

Clearly, if the proportion of symbols of the ith kind in the original (uncollapsed) sequence is π_i (where $i=1, \ldots, s$ and $\sum \pi_i=1$), then $q_{ij}=\pi_j/(1-\pi_i)$. Let us also write $\mathbf{b}'=(b_1, b_2, \ldots, b_s)$ for the limiting vector of the chain and $\mathbf{M}=\{m_{ij}\}$ for the matrix of digram frequencies.

To test the hypothesis it is desirable to sample several, say t, transects; each of them yields a collapsed sequence from which a matrix of digram frequencies can be compiled. Then, before testing whether the collapsed sequences can be regarded as realizations of a simple Markov chain, a preliminary test should be done to judge whether the observed sequences are homogeneous. That is, assuming each observed sequence forms a simple Markov chain, we test the conditional null hypothesis that all of them have the same matrix of transition probabilities, in other words, that the mosaic under investigation is stationary. This preliminary test is carried out as follows (Kullback, Kupperman and Ku, 1962).

Let the frequency in the (i, j)th cell of the kth digram matrix be m_{ijk}. Put

$$\sum_i m_{ijk}=m_{\cdot jk} \qquad \sum_k m_{ijk}=m_{ij\cdot} \qquad \sum_i\sum_k m_{ijk}=m_{\cdot j\cdot}.$$

Then the log likelihood ratio required for the test is

$$-\ln \lambda = \sum_i \sum_j \sum_k m_{ijk} \ln m_{ijk} - \sum_j \sum_k m_{\cdot jk} \ln m_{\cdot jk} - \sum_i \sum_j m_{ij\cdot} \ln m_{ij\cdot} + \sum_j m_{\cdot j\cdot} \ln m_{\cdot j\cdot}.$$

Under the conditional null hypothesis, $-2 \ln \lambda$ is a χ^2 variate with $s(s-2)(t-1)$ degrees of freedom.

Table 5.3 Test of the homogeneity of two matrices of digram frequencies, on the assumption that they represent transition frequencies in simple Markov chains. The phases are labeled W, X, Y, and Z.[a]

$\mathbf{M}_1 =$

	W	X	Y	Z
W	0	9	20	9
X	10	0	5	7
Y	18	6	0	5
Z	9	7	4	0

$\mathbf{M}_2 =$

	W	X	Y	Z
W	0	18	13	9
X	18	0	6	6
Y	14	5	0	2
Z	7	7	3	0

$s=4; \quad t=2$

$-2 \ln \lambda = 6.537$

$P(\chi^2 > 6.537 \mid 8) > 0.5$

[a] From Pielou, 1967.

A numerical example (from Pielou, 1967) is shown in Table 5.3.

If this preliminary test gives no reason to doubt that the mosaic is homogeneous, the t digram matrices can be added together and tested for Markovity. A recipe for the test is given below and Table 5.4 gives an illustrative example using artificial data. These data were generated by taking digits from a random numbers table and assigning the symbol A to 0, 1, 2, and 3; B to 4, 5, and 6; C to 7 and 8; D to 9; and then collapsing the sequence. The parameters π_i, b_i and q_{ij} (with $i, j = 1, \ldots, 4$) are thus known *a priori* but, in what follows, they have been estimated from the "observations" as is necessary in practice. (All derivations, and an application of the test to real data are given in Pielou, 1967.) The steps are as follows.

1. Take the observed proportions of the A's, B's, ... in the collapsed sequence (as given by the row totals of the diagram matrix) as estimates of the elements of $\mathbf{b}'$.

2. The π's can be obtained from the b's using the relation

$$\pi_i = \frac{1-\sqrt{1-4Cb_i}}{2} \tag{5.4}$$

where C is a constant of proportionality whose value must be such as to ensure that $\sum \pi_i = 1$. Find C by successive approximation. Clearly we must have $C < 1/4b_1$ where $b_1 = \max(b_i)$.

3. Evaluate the elements of $\boldsymbol{\pi} = (\pi_1, \pi_2, \ldots, \pi_s)$ from (5.4).
4. Evaluate the expected digram frequencies from

$$m_{ij} = Mb_i q_{ij} = \frac{Mb_i \pi_j}{1 - \pi_i}$$

where

$$M = \sum \sum m_{ij}.$$

Observe that $\mathscr{E}(\mathbf{M})$ is symmetrical and its row totals are identical with the observed row totals.

Table 5.4 To test that the collapsed sequence that yielded the digram matrix **M** is a realization of a simple Markov chain. The phases are A, B, C and D.

M ("observed") =

	A	B	C	D
A	0	49	28	13
B	48	0	16	14
C	27	18	0	5
D	16	10	6	0

$M = 250$

$\mathbf{b}' = (0.360 \quad 0.312 \quad 0.200 \quad 0.128)$

$C = 0.682065$

$\hat{\boldsymbol{\pi}}' = (0.43324 \quad 0.30714 \quad 0.16297 \quad 0.09665)$

$\mathscr{E}(\mathbf{M})$ =

	A	B	C	D
A	0	48.77	25.88	15.35
B	48.77	0	18.35	10.88
C	25.88	18.35	0	5.77
D	15.35	10.88	5.77	0

$X^2 = 2.008$; $\quad -2 \ln \lambda = 2.328$; $\quad \nu = 7$;

$P(\chi^2 > 2.328 \mid 7) > 0.9.$

5. Do a goodness-of-fit test calculating as test statistic either

$$X^2 = \sum \frac{(O-E)^2}{E}$$

or

$$-\ln \lambda = \sum \sum m_{ij} \ln m_{ij} - \sum m_{i\cdot} \ln m_{i\cdot} - \sum \sum m_{ij} \ln q_{ij}.$$

Both X^2 and $-2 \ln \lambda$ are asymptotically distributed as χ^2 with $\nu = s - 2s - 1$ degrees of freedom.

It seems safe to predict that studies of spatial patterns in stationary mosaics will contribute greatly to an understanding of community diversity. No one can yet say whether, in such mosaics, random mingling is the rule or the exception. In cyclical mosaics, in mosaics containing allelopathic species of plants, and in mosaics in which the presence of some species modifies the habitat for others, random mingling would not be expected. It is not yet known whether mosaics of these kinds are commonplace or unusual; even if conspicuous examples are uncommon, less obvious examples may abound.

5.4 The Number of Species per Sampling Unit

This section deals with two tests to judge whether species that occur in a community as separate individuals, rather than as patches, are randomly mingled.

The first of the tests entails comparisons between observed and expected distributions of the variate: number of species per sampling unit (s.u.), say z. The s.u.'s may be arbitrary, such as quadrats; or naturally discrete pieces of habitat such as rock pools, caves, rotting logs, bodies (as parasite habitats), and so on. The test is due to Barton and David (1959).

Suppose there are s species and the ith is present in n_i of a sample of N s.u.'s examined, or equivalently in a proportion $p_i = n_i/N$ of them. Then the null hypothesis is, that for $i = 1, \ldots, s$, every s.u. has the same probability, namely p_i, of containing species i. Counting the number of species, z, in a single s.u. is thus equivalent to counting the number of "successes" in s independent Bernoulli trials whose probabilities of success are $p_1, p_2, \ldots, p_s$. Consequently, the probability generating function (pgf) of z, under the null hypothesis, is

$$f(\theta) = \prod_{i=1}^{s} (q_i + p_i\theta) \tag{5.5}$$

where $q_i = 1 - p_i$.

Barton and David show that (provided the p_i values do not vary too greatly) this distribution may be closely approximated by a binomial distribution having the same mean and variance. Let this approximating binomial have pgf $(Q+P\theta)^S$ so that its mean and variance are, respectively, SP and SPQ. We now require estimates of S, P, and $Q(=1-P)$ in terms of the observed quantities s, N, and the p_i values.

The mean, $\mathscr{E}(z)$, and variance, Var (z), of the distribution whose pgf is given in (5.5) are derived as follows.

Denote the observed mean and variance of the s values of p_i by $\bar{p}$ and σ_p^2.

Then

$$\mathscr{E}(z)=\sum_{i=1}^{s} p_i = s\bar{p};$$

and

$$\begin{aligned}\text{Var}\,(z) &= \sum_{i=1}^{s} p_i q_i = \sum (p_i - p_i^2)\\ &= \sum \{(p_i - 2p_i\bar{p} + \bar{p}^2) - (p_i^2 - 2p_i\bar{p} + \bar{p}^2)\}\\ &= s\bar{p}(1-\bar{p}) - s\sigma_p^2.\end{aligned}$$

Now equate these moments with those of the approximating binomial. That is, put

$$s\bar{p} = SP \qquad \text{and} \qquad s\bar{p}(1-\bar{p}) - s\sigma_p^2 = SPQ.$$

Solving for P and S yields

$$P = \bar{p} + \frac{\sigma_p^2}{\bar{p}} \qquad \text{and} \qquad S = s\left[1 + \frac{\sigma_p^2}{\bar{p}^2}\right]^{-1}$$

as the estimators of the parameters of the approximating binomial.

As an example, consider the data given by Culver (1970) on the occurrences of $s=7$ species of cave animals (three amphipods, an isopod, two crayfishes, and a salamander) in $N=28$ limestone caves (the s.u.'s) in West Virginia. As Table 5.5 shows, the observed distribution of the number of species per cave appears to be binomial.

Clearly, observed distributions will depart from expectation if, for all i, p_i is not constant for every s.u. as the null hypothesis assumes. The form of the departure will depend on whether the p_i values tend to be negatively or positively correlated among themselves.

If the p_i values are negatively correlated, the observed number of empty s.u.'s (for which $z=0$) will fall short of expectation leading one to infer

Table 5.5 The fit of a binomial distribution to the observed distribution of the number of species of animals in limestone caves reported by Culver (1970)

z	Frequency: Observed	Frequency: Expected
0	3 }	8.0
1	6 }	
2	8	9.1
3	5	7.1
4	4 }	4.8
≥5	2 }	
	28	

$$X^2 = \sum \frac{(O-E)^2}{E} = 2.153$$

$$P(\chi^2 > 2.153 \mid 3) > 0.5$$

$N = 28$

$\{n_i\} = \{Np_i\} = (16 \quad 11 \quad 10 \quad 8 \quad 7 \quad 6 \quad 5)$

$\bar{p} = 0.3214 \qquad \sigma_p^2 = 0.01531$

Therefore the appropriate parameters for the fitted binomial distribution are

$$P = 0.37 \quad \text{and} \quad S \simeq 6.$$

either that the s.u.'s differ among themselves so that some provide optimal habitats for some species and others for others; or that the different species tend to exclude one another from places they have preempted. In either case, the species will be spatially segregated rather than randomly mingled.

Alternatively—and this appears at the moment to be the commoner phenomenon—the p_i values will be positively correlated. Then the number of empty s.u.'s will exceed expectation. This can be described as a manifestation of the "water hole effect", that is, the tendency for most or all of the species in a community to congregate (actively or passively) in places favorable to all of them. Then an s.u. that coincides with, or includes, a favorable place will contain a large number of species and an s.u. that does not will be empty. As a result, what has been called "point diversity" (Slobodkin and Fishelson, 1974) will vary from place to place. Examples are numerous. One of the most obvious occurs in semiarid country (such as East African savanna) where many animals, of which large ungulates may be the most visible, assemble at water holes at certain times of the day. Another example has been given by Slobodkin and Fishelson who describe the way in which reef fishes of many species congregate around the "stations" of cleaner-fish (a wrasse). A third example, probably also the result of a behavioral response of animals to sites of varying attractiveness, has been given by Pielou and Verma (1968); it was found that bracket fungi varied greatly in the number of species of immature arthropods they contained.

Point diversity can also vary passively. Suppose species that are preyed upon by carnivores live in a spatially variable habitat in which the cover is good at some sites and poor at others. Then, even if the prey populations do not actively select places with good cover they will at least survive there, whereas those in poorly protected places will be captured more easily. Thus point diversity will be highest where concealment is most complete (Smith, 1972).

The second test to be considered in this section is applicable to communities of sessile organisms that, like forest trees, exist as discrete individuals rather than as extended patches. It entails comparison between observed and expected distributions of the variate: number of species per group of n, say, neighboring individuals. Whereas in the test described above no restriction was placed on the number of individuals in a sampling unit, we now allow the size of the s.u.'s to vary so that each contains exactly n individuals. These might be the n individuals nearest to a random point; or the nearest individuals in n equal sectors centered on a random point; or any other convenient "plotless sample" of n individuals. For convenience, the entity sampled will be called a "group." If the composition of the whole community is completely known, it is easy to derive the expected distribution of x, the number of species, per group; obviously p_x, the probability that a group contains x species is a sum of s-dimensional hypergeometric probabilities; (s is the number of species in the community).

For example, suppose a complete census has shown the community to contain a total of N individuals with N_i in the ith species ($i=1, \ldots, s$). The probabilities p_x for $n=3$ and 4 are then as tabulated below:

n	x	p_x
3	1	$\dfrac{\sum_i N_i^{(3)}}{N^{(3)}}$
	2	$\dfrac{3\sum_{i \neq j} N_i^{(2)} N_j}{N^{(3)}}$
	3	$\dfrac{6\sum_{i<j<k} N_i N_j N_k}{N^{(3)}}$

$$4 \left\{ \begin{array}{ll} 1 & \dfrac{\sum_{i} N_i^{(4)}}{N^{(4)}} \\[2ex] 2 & \dfrac{\left\{4 \sum_{i \neq j} N_i^{(3)} N_j + 6 \sum_{i<j} N_i^{(2)} N_j^{(2)}\right\}}{N^{(4)}} \\[2ex] 3 & \dfrac{12 \sum_{i \neq j,k \; j<k} N_i^{(2)} N_j N_k}{N^{(4)}} \\[2ex] 4 & \dfrac{24 \sum_{i<j<k<l} N_i N_j N_k N_l}{N^{(4)}} \end{array} \right.$$

where $N^{(r)} = N(N-1) \cdots (N-r+1)$.

It is straightforward, but laborious, to extend the table to include higher values of n. It is also straightforward to calculate a diversity index for each group sampled but this adds little to the usable information obtained. If the species are spatially segregated, low values of x will be more frequent than expected; and if they are negatively segregated (more thoroughly mingled than would happen by chance) high values of x will be more frequent than expected. Therefore, if one wishes to test the null hypothesis that the mingling is random, an ordinary goodness-of-fit test is all that is required.

Isolated tests of this kind are seldom very instructive, of course. For results to be interpretable, comparisons are necessary. For example, Pielou (1966*a*) examined the way in which spatial segregation changed with time in even-aged tree communities as they grew older. Even-aged second-growth forest undergoes natural thinning as it ages, since the seedling trees are far too dense initially for all to survive to maturity. If most deaths occur in dense one-species clumps of seedlings where seeds have germinated around a parent tree, thinning will cause a progressive reduction in spatial segregation of the species (an increase in the degree of mingling). On the other hand, if the habitat is nonuniform and the seedlings tend to survive only in places to which they are adapted, and to die off if they chance to have germinated in suboptimal places, then thinning will be accompanied by an increase in spatial segregation. In the forest plots investigated, segregation appeared to decrease with time. The conclusion, that natural thinning has the effect of reducing the density chiefly in dense one-species clumps implies also that sparse, uncrowded species will be comparatively unscathed and hence that overall diversity will not diminish as a forest matures.

Chapter 6

Diversity on Environmental Gradients

6.1 Introduction

In a community whose spatial pattern is stationary (as defined in Section 5.2), exogenous and endogenous contributions to diversity are difficult to distinguish. There are two ways of attempting to overcome the difficulty, one practicable and one not. The latter is to study communities in "absolutely uniform" environments, but these are so rare, and their putative uniformity so dubious, that they are unpromising objects of study. The feasible solution to the difficulty is to search for communities whose pattern is governed by monotonic unidirectional variation in some one environmental factor of overwhelming importance to the organisms concerned; that is, to study *zoned communities* on an environmental gradient.

An interest in environmental gradients and their ecological effects has inspired many investigations on a wide array of different plant and animal groups. To mention only a few examples, zonation has been studied in: the trees of mountain forests (Whittaker, 1972), the birds of mountain forests (Diamond, 1973), intertidal marine algae (Chapman, 1973), barnacles (Connell, 1961), and salt marsh vegetation (Pielou, 1975*b*).

The questions asked and the methods for attempting to answer them are as diverse as the organisms. Indeed, zonation studies are a paradigm of community research for they illustrate clearly the way in which two contrasted approaches can be made to the same set of problems. One approach consists in constructing simple theoretical models of species interactions, deriving their consequences, and comparing these with real data; the other consists in deciding what field data need to be collected, and how these data should be analysed, to yield unambiguous answers to specific questions. The following sections illustrate both these approaches.

6.2 Modeling Competition on a Gradient

Consider a many-species community of sessile organisms occurring on an environmental gradient. If it were not for interspecific competition, each species would occupy a zone whose position and width would depend on that species' tolerance range for the gradient variable. However, because of competitive exclusion, many of the species are likely to be confined to relatively narrow zones. We shall use the terms *fundamental zone* and *realized zone* respectively (cf Section 8.3) to denote the zone a species is physiologically capable of occupying and the zone in which it succeeds in establishing itself.

Now consider how the interactions among species on an environmental gradient might be modeled mathematically. The simplest of all models to deal with in a uniform environment is that described by Gause's equations:

$$\frac{1}{A}\frac{dA}{dt}=r_A-s_AA-u_AB$$
$$\frac{1}{B}\frac{dB}{dt}=r_B-s_BB-u_BA. \tag{6.1}$$

Here it is assumed that only two species, A and B, are present; that the *per capita* growth rate of each species-population is a linear function of the number of competitors present of the same and of the other species; that growth rates respond without delay to the changing sizes of the competing populations; that conditions remain constant; and that stochastic effects can be ignored. The population sizes are A and B, and the parameters are to be interpreted as follows; r is a population's intrinsic rate of increase; s is the competition coefficient measuring a species-population's inhibitory effect on its own growth rate; u is the coefficient measuring its inhibitory effect on the other population's growth rate. As is well known:

if $r_A u_B > r_B s_A$ and $r_A s_B > r_B u_A$,
species A will exclude species B;

if $r_B s_A > r_A u_B$ and $r_B u_A > r_A s_B$,
species B will exclude species A;

if $r_A s_B > r_B u_A$ and $r_B s_A > r_A u_B$,
both species can coexist in stable equilibrium;

and if $r_A u_B > r_B s_A$ and $r_B u_A > r_A s_B$ (the "unstable equilibrium" case), one species is bound to become extinct leaving the other as sole survivor but which species will win depends on the initial values given to A and B.

Imagine that populations of species A and B colonize an environmental gradient and let G denote the value of the gradient variable. Assume that (6.1) describes population growth at any point on the gradient and that the parameters are functions of G. Suppose, in particular, that the parameters vary in such a way that, in a three-dimensional coordinate frame with axes A, B, and G, the isoclinal surfaces $(1/A)dA/dt=0$ and $(1/B)dB/dt=0$ are planes. Then three-dimensional figures such as those in Figure 6.1 represent the model diagramatically. The direction of the gradient is perpendicular to the plane of the page; and the two-demensional cross-sections in the plane of the page are phase-space graphs showing the trajectories (curved arrows) of mixed populations of A's and B's for some one value of G. Each diagram in the figure represents conditions in part of a transition zone on the gradient, between a zone occupied exclusively by species A and a zone occupied exclusively by species B. But, as is obvious from the diagram, two different kinds of transition zone are possible: an unstable zone in which the two species cannot coexist, and a stable zone in which they can.

Now visualize an environmental gradient in nature supporting a two-species community of A's and B's; at one end of the gradient is a zone where A occurs alone and at the other end a zone where B occurs alone. If (6.1) describes the growth rates of the two species-populations, one would expect (intuitively) that if the transition zone were stable, the two "pure" zones would blend into each other to give a mixed zone in which the two species coexisted in stable equilibrium; and that if the transition zone were unstable, the two pure zones would meet along an abrupt boundary with no blending.

This latter expectation (of an abrupt boundary in the unstable case) goes somewhat beyond what the model predicts. Recall that the model assumes an enclosed space and that, for given values of the parameters, the outcome (total exclusion of one species by the other) depends only on the

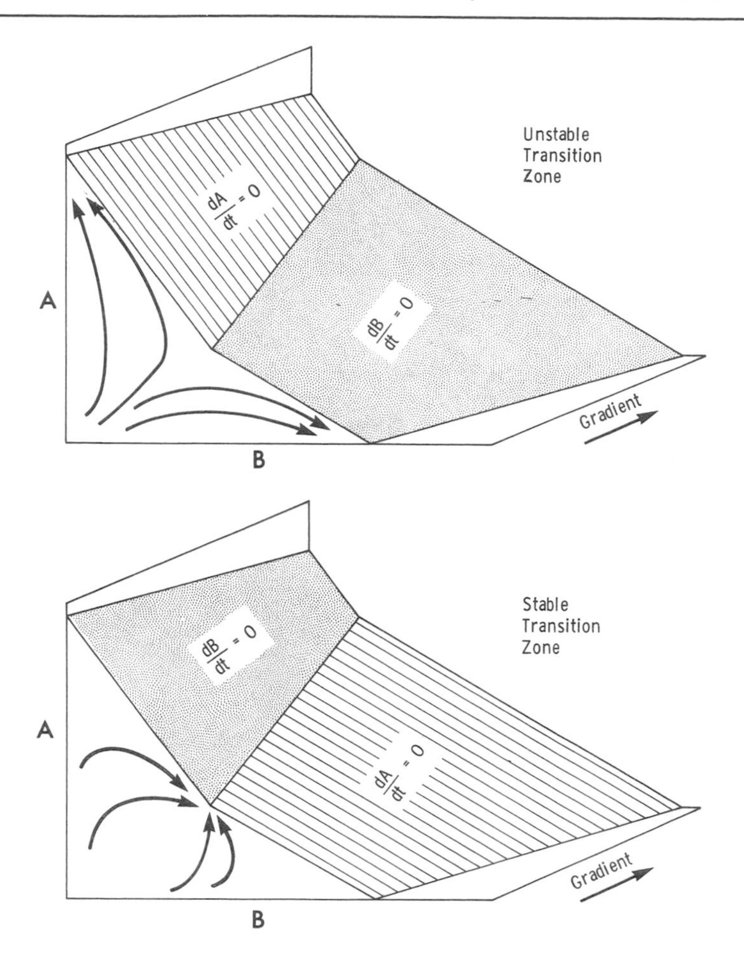

Figure 6.1 Two-species competition on an environmental gradient. The isoclinal planes are $dA/dt = 0$ (hatched) and $dB/dt = 0$ (stippled).

number of members of each species present initially, whatever their location in the enclosed space. In a zoned community on a gradient, the only uniform environments are vanishingly narrow strips, and they are certainly not closed to immigration. If the model is to make the predictions required, therefore, it must include the stipulation that everywhere on A's side of a boundary line bisecting the unstable transition zone, mixed starting populations will always contain a sufficiently high proportion of A's to ensure that A will win and B become extinct; and conversely on B's side of the boundary.

We now have a model that predicts a blended zone between two one-species zones in some cases, and an abrupt boundary between contiguous pure zones in other cases. It could explain actual realizations of these phenomena; thus in salt marshes, as an example, the zones are usually blended, whereas intertidal attached algae on sheltered rocky shores usually show abrupt zonation. In spite of the model's unrealistic simplifications, it could account for the salient fact that zones sometimes do, and sometimes do not, blend. However we cannot judge whether it does or not unless other predictions can be derived from the model which can also be compared with processes in natural populations. To make the model yield additional predictions it may conveniently be modified by substituting difference equations for differential equations, by treating separately the separate age classes within each species-population, and by allowing the effects of within- and between-species competition to depend on age (Pielou, 1974*a*).

The changes in the model are as follows. Consider, first, events at any one given level of the gradient. Equations 6.1 are replaced by their equivalent difference equations, namely

$$A(t+1)=\frac{\lambda_A A(t)}{1+\alpha_A A(t)+\gamma_A B(t)}$$
$$B(t+1)=\frac{\lambda_B B(t)}{1+\alpha_B B(t)+\gamma_B A(t)} \qquad (6.2)$$

where $A(t)$ and $B(t)$ denote the sizes of the competing populations at time t, and where (subscripting throughout with A or B as required):

$$\lambda=e^r; \qquad \alpha=\frac{s(\lambda-1)}{r}; \qquad \gamma=\frac{u(\lambda-1)}{r}.$$

The equivalence of (6.2) and (6.1) is demonstrated in Leslie (1958) and Pielou (1969).

Now take account of the age structures of the two species-populations. Assume that the maximum attainable age for species A is k time units, and for species B is l time units, and let the whole two-species community at time t be represented by the vector $\mathbf{v}(t)$ where

$$\mathbf{v}'(t)=(a_{0t}, a_{1t}, \ldots, a_{kt}, b_{0t}, b_{1t}, \ldots, b_{lt}).$$

Here a_{xt} is the number of members of species A aged x at time t, and correspondingly for b_{xt}. Thus $\sum_{x=0}^{k} a_{xt}=A(t)$ and $\sum_{x=0}^{l} b_{xt}=B(t)$.

Then the growth of the whole community can be modeled by an

equation of the form

$$\mathbf{v}(t+1)=\mathbf{Q}(t)\,\mathbf{v}(t) \tag{6.3}$$

where $\mathbf{Q}(t)$ is a *projection matrix* (or "Leslie matrix") defined by

$$\mathbf{Q}(t)=\left(\begin{array}{ccccc|ccccc} f_0 & f_1 & \cdots & f_{k-1} & f_k & & & & & \\ p_0 & 0 & \cdots & 0 & 0 & & & \mathbf{0} & & \\ \cdot & \cdot & \cdot & \cdot & \cdot & & & & & \\ 0 & 0 & \cdots & p_{k-1} & 0 & & & & & \\ \hline & & & & & F_0 & F_1 & \cdots & F_{l-1} & F_l \\ & & \mathbf{0} & & & P_0 & 0 & \cdots & 0 & 0 \\ & & & & & \cdot & \cdot & \cdot & \cdot & \cdot \\ & & & & & 0 & 0 & \cdots & P_{l-1} & 0 \end{array}\right)$$

Here f_j is the number of offspring born per unit of time to an A-individual aged j that survive into the next time interval; p_j is the proportion of A-individuals aged j that survive to age $j+1$; F_j and P_j are the corresponding parameters for species B. $\mathbf{Q}$ is a function of t since it is assumed that the f's, p's, F's, and P's at time t depend on the current sizes of the competing populations. They are defined as

$$f_j=\frac{\phi_j}{1+\alpha'_{Aj}A(t)+\gamma'_{Aj}B(t)}$$

$$p_j=\frac{\theta_j}{1+\alpha_{Aj}A(t)+\gamma_{Aj}B(t)} \qquad \text{with } j=0,1,\ldots,k;$$

and (6.4)

$$F_j=\frac{\Phi_j}{1+\alpha'_{Bj}B(t)+\gamma'_{Bj}A(t)}$$

$$P_j=\frac{\Theta_j}{1+\alpha_{Bj}B(t)+\gamma_{Bj}A(t)} \qquad \text{with } j=0,1,\ldots,l.$$

The α''s and γ''s measure the decrease in fertility and the α's and γ's the decrease in survival, caused by within- and between-species competition respectively.

If we now allow all parameters to be functions of G, the gradient level, an improved version (that is, one allowing for the effects of age-structure) of the simple model portrayed graphically in Figure 6.1 results. It will be noted that even in a uniform environment (with no gradient) the number of parameters that would need to be specified to run a simulation of the model is $6(k+l+2)$ since for every age class of each species there are 6 parameters in need of numerical values on the right-hand side of (6.4).

A simple version of the model has been simulated (Pielou, 1974*a*) and led to the following conclusion; (whether it can safely be generalized is not known). When an area with a strong environmental gradient is colonized by two competing species, the age distribution of each species in the transition zone soon matches that in the corresponding one-species zone if the former zone is stable and hence will permanently support a "blended" mixture of the species at equilibrium. But if the transition zone is unstable, then in the mixed population of A's and B's occupying it before equilibrium is attained, the proportion of young individuals in the temporary (losing) subpopulation of A's on B's side of the ultimate boundary is higher than in the permanent (winning) subpopulation of A's on A's own side of the boundary; and vice versa for the temporary and permanent subpopulations of B's.

Whether realizations of this simulated phenomenon will be found in natural communities remains to be discovered. In any case, observations matching these predictions would not necessarily imply the correctness of the model. Many other causes might well produce the same result and it is worth considering how the model would behave if two of its oversimplified assumptions were relaxed.

First, recall the assumption that, once equilibrium has been reached, each species will be found only in its own zone and in the transition zone if the latter is stable. This seems unrealistic. It seems more likely that in practice there will often be a spillover effect, with some individuals of each species appearing sporadically and temporarily in the wrong zone. In zoned vegetation, for example, when the death of an established plant leaves a gap to be colonized, it is not surprising to find it invaded by a species from a neighboring zone even though the intruder will not survive for long nor leave offspring around itself. If this argument is correct then two different kinds of mixed zones are possible: *blended zones* in which species coexist in equilibrium; these are stable transition zones hypothesized above. And *blurred zones* in which species coexist in disequilibrium; because of the permanent close proximity, as sources of immigrants, of neighboring equilibrium zones (either pure or blended), blurred zones can persist in spite of disequilibrium.

We now enquire whether it is possible to distinguish between blending and blurring in the field. A way of recognizing blurring is as follows.* Visualize a map of a zoned community (see Figure 6.2) showing isopleths of the gradient variable; (if the gradient variable is very strongly correlated with altitude, as it is for the zoned communities of mountainsides and shores, these isopleths are ordinary topographic contours). Then if there is

* The idea is due to Richard Routledge.

no blurring the zone where a species occurs, either alone or in stable coexistence with others, will have isopleths as its boundaries; that is, its width with respect to the gradient variable, or "vertical width," will be constant. But if the boundaries of a species' zone are blurred, because the region in which a species can maintain itself in equilibrium (alone or with other species) is bordered by strips where disequilibrium subpopulations are endlessly replenished, then the zone in which the species occurs will not be of constant vertical width; its on-the-ground width will tend to be less variable than the widths of strips bounded by contours, so that its vertical width will tend to be somewhat greater on steep slopes than on gentle ones. Evidence so far available shows that in salt marsh vegetation this does indeed happen (Routledge, pers. comm.).

The problem of judging whether a mixed zone consists only of a blurred zone, or of a blended zone bordered by blurred zones, is more difficult. In theory one could judge the relative importance of the blending and blurring mechanisms from examination of a map like that in Figure 6.2. In practice this would obviously be difficult. It is certainly not inconceivable, however, that mixed zones in zoned communities *always* result from blurring rather than blending; or, equivalently, that when several species occur in the same zone, all but one are always in disequilibrium.

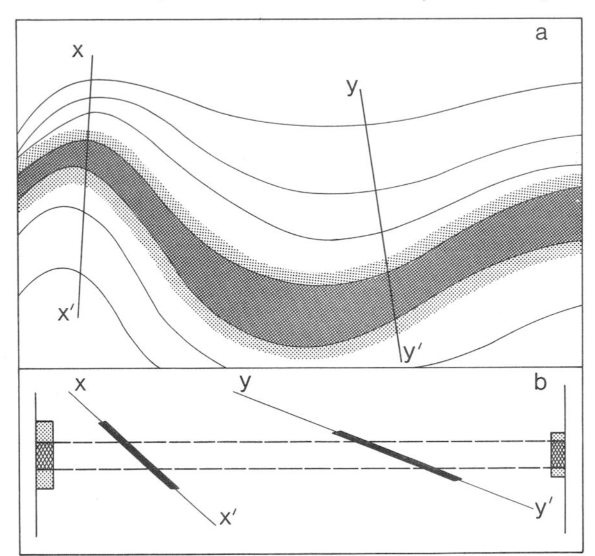

Figure 6.2 (*a*) Map showing the zone occupied by one species on a slope; the species occurs at equilibrium in a zone (crosshatched) bounded by contours, and in disequilibrium in bordering strips (stippled) of constant width on the ground. (*b*) Sections through XX' and YY'.

Consider now the effect of relaxing another of the original model's oversimplified assumptions. We use (6.2) as a starting point and, as before, allow the parameters to vary with the gradient level. No allowance will be made for the effects of the populations' age structures, nor for the "blurring" of zone boundaries discussed above. We shall, however, allow for delayed response of each species' growth rate to competition pressure by replacing (6.2) with the following equations of growth:

$$A(t+1)=\frac{\lambda_A A(t)}{1+\alpha_A A(t-d_A)+\gamma_A B(t-d_A)}$$
$$B(t+1)=\frac{\lambda_B B(t)}{1+\alpha_B B(t-d_B)+\gamma_B A(t-d_B)}. \tag{6.5}$$

Thus it is assumed that species A's growth rate at any time is controlled by the sizes of the two competing populations at a time $(t-d_A)$ time units earlier; and similarly for species B. If we now allow the d's as well as the λ's, α's, and γ's to vary with gradient level, and allow an area with an environmental gradient to be colonized by the two species, simulations show that the following events can happen (Pielou, 1974*b*): "repeated zones" will appear; that is, two or more zones dominated by species A will alternate with zones dominated by species B; further, the zones will migrate up and down the gradient (see Figure 6.3). Indeed, a community of "migrating stripes" is formed which can be regarded as a realization, in an environment with a gradient (which is therefore a nonstationary environment) of an endogenous cyclical mosaic such as those described in

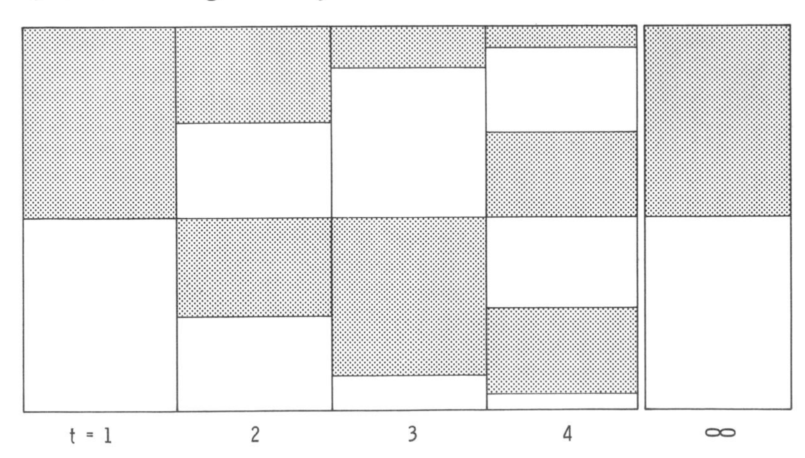

Figure 6.3 The pattern at a sequence of times of a mixed population of species A and species B growing in accordance with (6.5). The stippled and clear zones are dominated by A and B respectively, but total population sizes are not shown; (adopted from Pielou, 1974*b*).

Section 5.2. Ecological succession is a manifestation of delayed response and, on a gradient, stripes instead of isodiametric patches can form and migrate cyclically. At a given gradient level, the population sizes of the two species are oscillating and the oscillations are in different phases at different gradient levels. In the model shown diagramatically in Figure 6.3, the oscillations are damped so that ultimately a steady state is reached as shown on the right of the figure. It would be easy to devise a slightly more complicated model that would yield permanently migrating stripes; thus a model giving a stable limit cycle in a uniform environment (for examples, see May, 1973) would, on a gradient, give this effect.

It would also be easy, though perhaps unprofitable, to modify the original simple model of (6.1) and (6.2) in many other ways. It is worth recapitulating, in the form of a list, some of the model's simplifications that could (in theory) be allowed for, either singly or in combination:

1. Equations 6.1 assume the *per capita* growth rate of the two species to be linear functions of their population sizes. This is presumably unrealistic unless one is concerned with changes over a very small range of values only.
2. No allowance is made for stochastic effects, either demographic or environmental (May, 1973). Demographic stochasticity is the departure (because of chance) of the actual numbers of births and deaths from those predicted by a deterministic model; environmental stochasticity is the result of chance fluctuations, both spatial and temporal, in the values of the demographic parameters.
3. In bisexual species, the sex ratio could be allowed for, as well as the age structure within each sex.
4. More elaborate allowances could be made for delayed responses. Thus (6.5) assumes that species A's response lags behind both A's and B's population sizes to the same extent. And a distinction could be made between reproductive-time lag and reaction-time lag (Wangersky and Cunningham, 1956).
5. The most desirable, as well as the most difficult, modification is to allow for $s>2$ species. Strobeck (1973) has derived the necessary and sufficient conditions for $s>2$ species to coexist and has shown that equilibrium depends on the species' intrinsic rates of increase as well as the competition coefficients.

To modify the original model in such a way as to take care of all these (and other) complicating factors simultaneously and *realistically* is (in my view) a task whose probability of failure is so close to unity that it is not worth attempting. The great merit of simple models is that they show what *can* happen, and what curious (or, at least, nonobvious) phenomena are worth looking out for. The fact that the model in (6.5) yields a cyclical repetition of zones on a monotonic gradient is an example; as the model shows, it is possible for zone repetition to occur even on a gradient that is strictly monotonic; thus repeated zones do not automatically indicate a localized reversal of the gradient. This is the great value of simple mathematical models: they show what can happen, and it is sometimes surprising.

6.3 Species Turnover Along a Gradient, or "Beta Diversity"

As an alternative to constructing "working" models of zoned communities, we consider, in this and the two following sections, field observations and ways of interpreting them. Recall that in Section 6.2 a distinction was made between a species' fundamental zone and its realized zone. Competition will tend to make the realized zone narrower than the fundamental zone, whereas blurring will to some extent mask this effect.

In any case, the point diversity (using the term in its intuitive sense) at a point in a zoned community depends both on the number of species in the community as a whole, and on the widths of their realized zones. For a given number of species suppose that, relative to the down-gradient length of the community, these zones are wide; then they must overlap one another considerably. As a result, the point diversity will be high everywhere, but at the same time the rate of change of species-composition as one travels across the zones will be low. Conversely, if the zones are narrow, overlap will be slight and point diversity values will be low; but the rate of change of species-composition as one crosses the zones will be high. The rate of change of species-composition has been called *beta diversity* by Whittaker (1972, and references quoted therein); other terms are "species turnover rate," "species replacement rate," and "rate of biotic change." Whittaker proposes the following method for measuring it.

Let a row of sampling plots or quadrats, equidistantly spaced, be ranged along the gradient; use interquadrat distance as the unit of distance. Choose a suitable similarity coefficient for measuring the similarity in species-composition between two quadrats (see below). From field data, obtain the mean similarity coefficient for pairs of quadrats separated by 1, 2, . . . units of distance along the gradient. Graph similarity against distance and from the graph find the distance that must separate two quadrats to ensure that their expected similarity shall be half that of two quadrats at the same level on the gradient, that is, at zero distance apart; (quadrats at the same level would not be expected to have identical species-compositions, of course, because of sampling variation). Then β, the reciprocal of this distance—the distance required for a "half-change" in species-composition—is defined as the beta diversity of the zoned community.

Two suitable measures of the similarity between two quadrats, say quadrat X and quadrat Y, are the *coefficient of community*, CC, and the *percentage similarity*, PS. They are defined as follows.

$$CC = \frac{200 s_{XY}}{s_X + s_Y}$$

where s_X and s_Y are the numbers of species in quadrats X and Y and s_{XY} is the number common to both quadrats.

$$PS = 200 \sum_{i=1}^{s} \min(P_{iX}, P_{iY})$$

where P_{iX} and P_{iY} are the quantities of species i in quadrats X and Y as proportions of the quantity of all s species in the two quadrats combined.

Clearly,

$$0 \le CC \le 100 \quad \text{and} \quad 0 \le PS \le 100.$$

Only when the two quadrats have no species in common will $CC = PS = 0$. If the two quadrats have identical species lists then, regardless of species quantities, $CC = 100$. But for $PS = 100$, it is necessary that the two quadrats have the same species in identical amounts.

In what follows we shall write R_d for the mean similarity (whether measured by CC, PS or some other coefficient) between quadrats a distance d apart along the gradient. Suppose that a row of n quadrats has been sampled and that as an estimate of R_d (for $d = 1, \ldots, n-1$) we take the mean of the observed similarity between the ith and $(i+d)$th quadrats averaged over the values obtained with $i = 1, 2, \ldots, (n-d)$. It has been found empirically, in the communities so far studied, that if $\log R_d$ is plotted against d, the relationship appears to be linear; Figure 6.4*a* shows an example. When the relationship is linear, an estimate of beta diversity may be obtained graphically as shown in Figure 6.4*b*.

A straight line is fitted by eye to those of the plotted points that are more or less colinear and the line is extended to right and left so that $\log R_0$ and $\log R_n$ can be read from the graph.

Then

$$\frac{1/\beta}{n} = \frac{\log R_0 - \log R_0/2}{\log R_0 - \log R_n},$$

whence

$$\beta = \frac{\log R_0/R_n}{n \log 2}$$

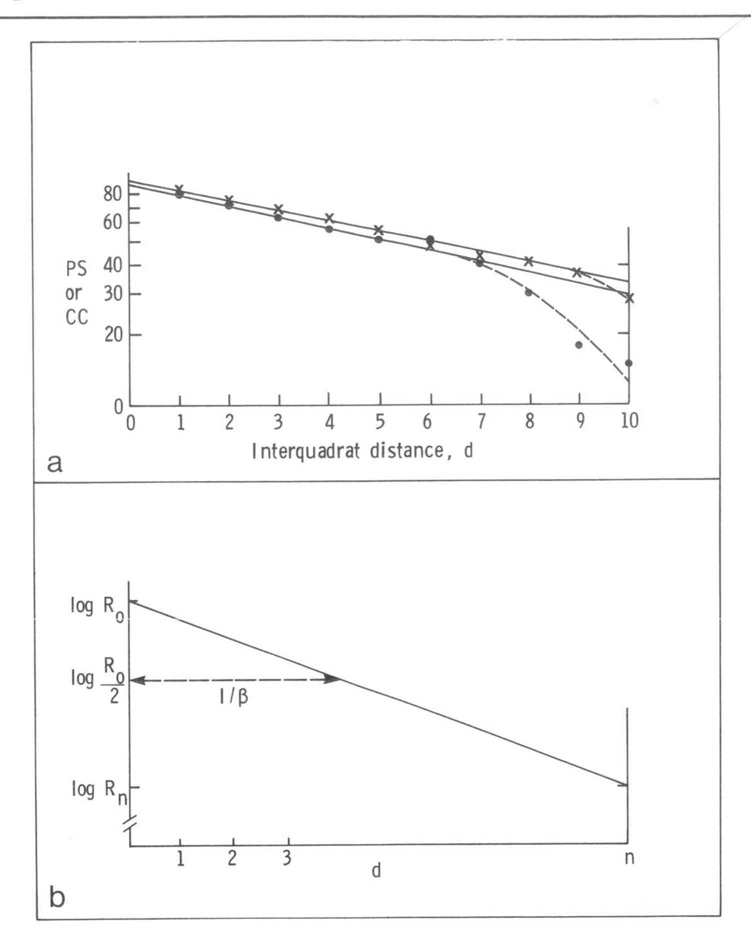

Figure 6.4 (*a*) The relationship between the similarity of vegetation samples and interquadrat distance along a topographic moisture gradient. The X's show values of *CC* and the 0's show values of *PS*. (Adapted from Whittaker, 1972.) (*b*) To illustrate the calculation of β.

is the required estimate of beta diversity. Thus, as stated above, β is the reciprocal of the distance, measured as a fraction of the length of the whole transect, between quadrat pairs whose mean similarity is $R_0/2$.

A method of estimating the sampling variance of β has not yet been derived. Nor is any allowance made for the possibility that, for given d, quadrat-pairs may tend to be more similar at one end of the gradient than the other. However, if these drawbacks can be remedied, comparisons among β values from different zoned communities will permit judgments as to how such communities differ in the widths and degrees of overlap of their zones.

6.4 Comparing Zone Patterns in Different Communities

It is particularly interesting to compare what may be called the *zone patterns* of zoned communities with many and with few species. The question at issue is: when many species occur together on a gradient, does competitive exclusion cause their realized zones to be narrower, relative to their fundamental zones, than when only a few species are present? Or are large groups of species able to coexist together at every level; that is, do wide, species-rich, blended zones occur? In studies of the altitudinal zonation of breeding birds on mountainous islands in the Pacific, Diamond (1973) found that on species-poor islands (with fewer than 66 species) congeneric species tended to have overlapping (blended) zones whereas in more species-rich islands zone boundaries were often abrupt.

A method of comparing the zone widths and overlaps of a given set of s species, say species A, B, C, ... that occur, with other species, in two different zoned communities is the following (Pielou, 1975*b*). It is necessary that each community be sampled along several transects. Let the number of transects, which should be the same for both communities, be k; k should be even. For each transect a list is made of the order of occurrence of the following events that occur as one travels from one end of the transect to the other, say down the gradient: the first encounter with each of the species A, B, C, ...; and the last encounter with each of these species. (All other species, in both communities, are disregarded.) Label these events with the symbols A_1, A_2, B_1, B_2, ... where the subscripts 1 and 2 denote the first and last encounter with the named species. For example the orders of the events in the two 3-species zone patterns shown in Figure 6.5 (going from the top downwards) are: in Figure 6.5*a*,

$$A_1\ B_1\ C_1\ A_2\ B_2\ C_2;$$

and in Figure 6.5*b*,

$$A_1\ A_2\ B_1\ B_2\ C_1\ C_2.$$

Assign the ranks 1, 2, ..., z (where $z = 2s$) to the events A_1, A_2, B_1, B_2, ... respectively. Then each transect list yields a list of the ranks of the z events and is thus a permutation of the numbers 1, ..., z. (For Figure 6.5*a* the list of ranks is 1, 3, 5, 2, 4, 6; for Figure 6.5*b* it is 1, 2, 3, 4, 5, 6.)

Now label the two communities X and Y and consider the k observed rankings (one for each transect) from one community, say X. Let R_j be the sum, over all k transects in this community, of the ranks of the jth event for

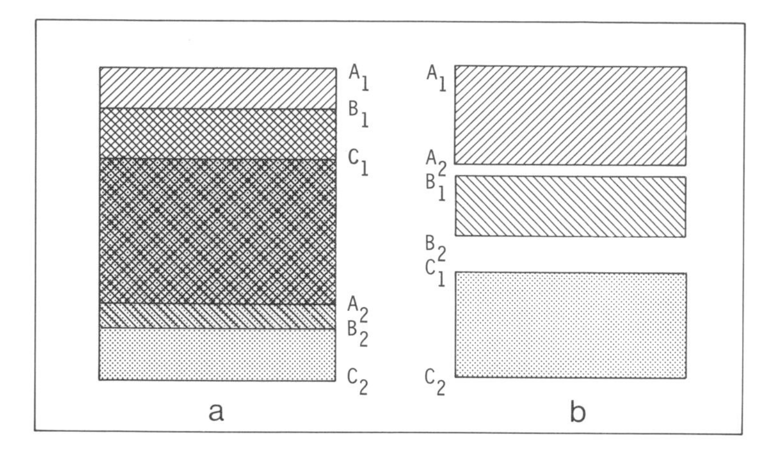

Figure 6.5 Two contrasting three-species zone patterns on a gradient. The gradient is up and down the page. In (*a*) the three zones overlap extensively; in (*b*) they do not. Other species assumed to be present are not shown.

$j=1, \ldots, z$. Then the *coefficient of concordance*, W_X, among the rankings from community X is defined as

$$W_X = \frac{12 \sum_{j=1}^{z} (R_j - \bar{R})^2}{k^2 z(z^2-1)}$$

where

$$\bar{R} = \frac{1}{z} \sum_{j=1}^{z} R_j = \frac{k(z+1)}{2}.$$

Thus W_X is the ratio of the observed variance of the R_j's to the variance they would have if all k rankings were identical, in which case the R_j's would be a permutation of the numbers $k, 2k, \ldots, zk$.

W_Y is defined analogously. Obviously $W_X, W_Y \leq 1$.

A coefficient of concordance (Kendall, 1948) is a measure of the closeness of agreement among several rankings and is usually calculated when the object is to test the null hypothesis that the rankings are unrelated, in which case the expectation of the coefficient is zero. In the present case, of course, the rankings are certainly not unrelated; also, some permutations of the ranks are impossible since the first encounter with a species necessarily precedes the last encounter. Therefore values of W_X and W_Y not far short of unity are to be expected. What we wish to determine is whether the correspondence among the rankings of all lists from both communities combined is significantly less than the correspondence within each community treated separately. In other words, is $W_{X+Y} < \min(W_X, W_Y)$? Here W_{X+Y} is the concordance among all $2k$ rankings from both communities.

Since the sampling variance of W when $\mathscr{E}(W) \neq 0$ is unknown, a direct test is impossible. A distribution-free test for deciding the question is as follows. Let $k=2m$. Pick, at random, m of the transect lists from community X, and m from community Y and combine (or "splice") them to give a new set of rankings whose coefficient of concordance among themselves, W_s say, may be calculated. Thus W_s is an estimate of between-community concordance which is based on the same number, $2m=k$, of rankings as W_X and W_Y, the estimates of the two within-community concordances. Further, since there are $\binom{k}{m}$ ways of selecting and splicing m transects from each community, $\binom{k}{m}$ independent values of W_s may be obtained. Thus $\binom{k}{m}$ independent comparisons between W_s and $\min(W_X, W_Y)$ are possible and if $W_s < \min(W_x, W_y)$ in a proportion α or less of the comparisons, then one can reject the hypothesis that the rankings, and hence the zone patterns, in the two communities are the same at the α level of significance. An example of the test is given in Pielou (1975*b*).

A practical point to notice is that in sampling along a gradient to record the first and last occurrences of several species, it is usually most convenient to examine a row of contiguous quadrats (equivalently, a belt transect divided into short sections) and compile a species list for each quadrat. The position along a transect of each event is then localized as to the quadrat in which it occurs rather than being observed exactly. Thus there are ties in the rankings when two or more species occur for the first or last times in the same quadrat. In calculating values of W, a correction for ties may be made as follows (Siegel, 1956). Assign to tied events the average of the ranks they would have been given if not tied. Let the number of events tied for a given rank in any one ranking be t; and let $T=\sum t(t^2-1)/12$, where the summation is over all groups of ties in all k rankings. Then the formula for W incorporating a correction for ties is

$$W=\frac{12\sum_{j=1}^{z}(R_j-\bar{R})^2}{k^2z(z^2-1)-kT}.$$

6.5 Further Analyses of Zone Patterns

All ecologists are familiar with the two opposed theories of the structure of plant communities: the *continuum concept* (otherwise known as the *individualistic concept*) and the *integrated community concept.* According to the former, the range of habitats (within its geographical range) in which a plant species grows depends only on the range of abiotic environmental factors to which the species is adapted, and species have all evolved independently. Therefore, where an environmental gradient causes zone formation, the boundaries of the realized zones will be located independently and at random (as in Figure 6.6*a*). According to the integrated community concept, groups of plant species tend to grow together; the members of a group are interdependent and for each species the presence of the others is as important an ingredient of the environment as are tolerable abiotic conditions. Therefore, on an environmental gradient, zone boundaries will tend to occur in batches (Figure 6.6*b*). A third concept is worth entertaining: that species have become so adapted to environmental conditions (whether abiotically or biotically determined) that, on a gradient, their zone boundaries tend to be evenly spaced out; shown diagramatically, as in Figure 6.6*c*, the zones overlap one another like tiles on a roof.

The zones shown in Figure 6.6 are, of course, realized zones; observation unaided by experiment cannot lead to incontrovertible conclusions about fundamental zones. However, a pattern such as that in Figure 6.6*d*, in which zones replace one another in such a way that the down-gradient boundaries of some tend to occur at the same level as the up-gradient boundaries of others, constitutes strong evidence for the occurrence of competitive exclusion and hence that the species' realized zones are (presumably) narrower than their fundamental zones. Wheras a pattern like that in Figure 6.6*e*, in which the locations of up-gradient and down-gradient zone boundaries are unrelated, suggests that competitive exclusion is unimportant; whether realized zones are narrower than fundamental zones nevertheless, it would be impossible to say.

Notice that in *a*, *b* and *c* of Figure 6.6, only one boundary, the up-gradient boundary of each zone is shown; down-gradient zone boundaries exhibit analogous patterns but there is no necessary correspondence between up-gradient and down-gradient boundary patterns in a given zoned community. Moreover the various combinations (there are obviously nine possibilities) of up-gradient and down-gradient boundary patterns can be put together as in *d* or *e* of the figure. Strictly, a pattern as in *d* can occur only if up- and down-gradient boundary patterns match but, as

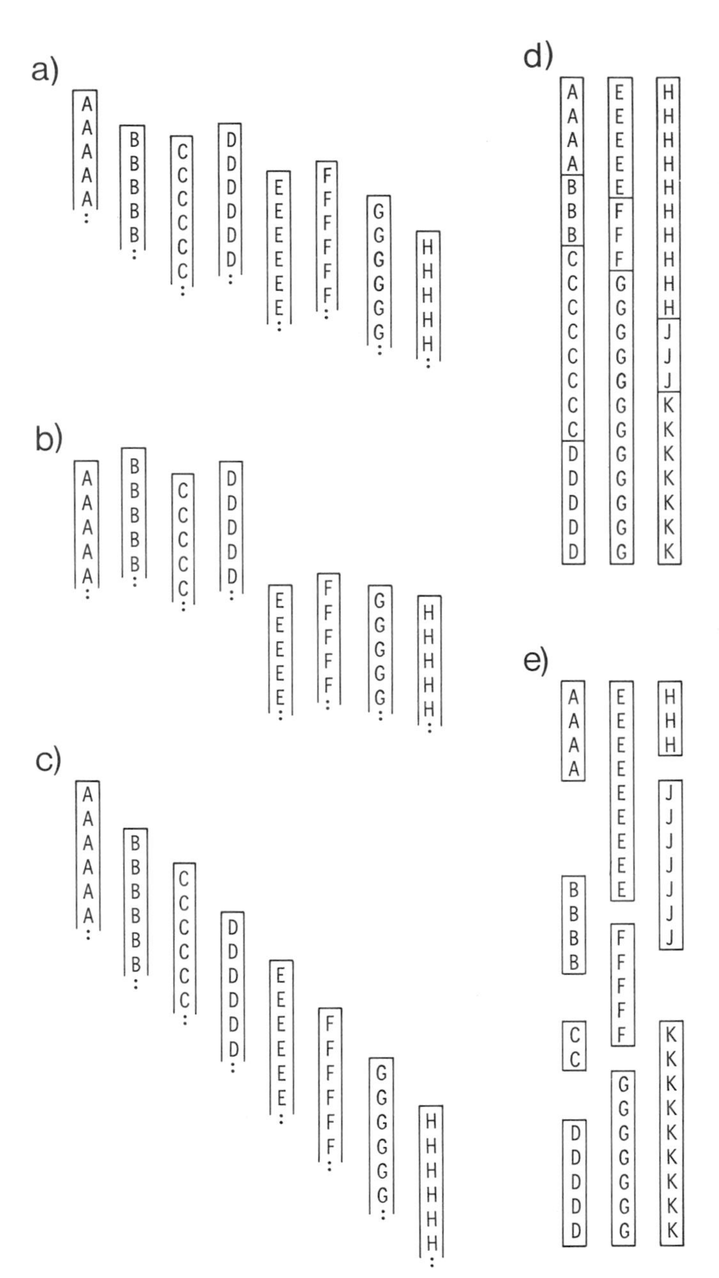

Figure 6.6 Possible zone patterns on a gradient. The gradient is up and down the page and the zones extend to left and right but for clarity they have not been shown superimposed. (*a*), (*b*), and (*c*) show possible patterns for the up-gradient boundaries of the zones. The down-gradient boundaries of the same zones might or might not have matching patterns. (*d*) and (*e*) show two ways in which zones may be interrelated.

we shall see below, assertions about what cannot happen may in practice prove false.

The ways in which these various possible patterns affect diversity, both point diversity anywhere along a gradient, and beta diversity, are interesting, if confusing, to contemplate. For example, point diversity at a sequence of points along a gradient would be expected to fluctuate erratically given certain types of zone patterns and vary smoothly given others. But more readily interpretable information on community structure can be had by studying zonation directly than by calculating diversities first. An example will illustrate this.

Suppose we wish to decide whether zone boundaries (both up- and down-gradient) in a zoned community have patterns as in *a*, *b* or *c* of Figure 6.6, that is, are at random, in batches, or regularly overlapped. Assume data are available from many transects across the community and that each transect consists of a row of contiguous quadrats for which complete species lists have been compiled. Let the number of quadrats in a given transect be Q; (Q will depend on the length and hence the steepness of the transect; it is not assumed to be the same for all transects). Let the total number of species encountered in the given transect be s. Now consider up-gradient zone boundaries and take as their estimated locations the locations of the "first encounters" with the species on going down-gradient from the "top" of the community. Take as null hypothesis that their pattern is random, as in Figure 6.6*a*. Under this hypothesis, the probability that x of the Q quadrats in a transect will *not* be cut through by an up-gradient zone boundary is

$$p_x = \frac{\left\{\begin{matrix}\text{Number of ways of}\\ \text{choosing } x \text{ of}\\ Q \text{ quadrats}\end{matrix}\right\} \times \left\{\begin{matrix}\text{Number of ways in which } Q-x \text{ quadrats}\\ \text{can contain } s \text{ boundaries with}\\ \text{none ``empty''}\end{matrix}\right\}}{\left\{\begin{matrix}\text{Number of ways in which } Q \text{ quadrats can}\\ \text{constain } s \text{ boundaries}\end{matrix}\right\}}$$

$$= \frac{\binom{Q}{x}\binom{s-1}{Q-x-1}}{\binom{Q+s-1}{s}}$$

with $(Q-s) \leq x \leq (Q-1)$.

By evaluating and summing these probabilities for given values of Q and s we may derive the expected median value of x, say $M(x)$. Thus if

$$\sum_{x=Q-s}^{M} p_x = 0.5 \qquad \text{then} \qquad M(x) = M;$$

and if

$$\sum_{x=Q-s}^{M} p_x < 0.5 < \sum_{x=Q-s}^{M+1} p_x, \qquad \text{then} \qquad M(x) = M + \frac{1}{2}.$$

When observed values of x, s, and Q are available from a large number of transects, it becomes possible to judge whether the observed frequency of the event $x < M(x)$ say, is improbably low, improbably high, or in accordance with expectation. The conclusions to be drawn would be, respectively, that the up-gradient zone boundaries were regularly overlapped, in batches, or at random. Analogously, one can diagnose the pattern of down-gradient boundaries from observations on "last encounters."

Examples of some results are shown in Table 6.1. These have been selected from data relating to 40 transects across two fairly similar salt marsh communities in Nova Scotia. It was found (Pielou, 1975*b*) that, for first encounters $x < M(x)$ in 22 of 40 transects, and for last encounters $x < M(x)$ in 30 of 40 transects. Under the null hypothesis, $\Pr\{x < M(x)\} = \Pr\{x > M(x)\} = 0.5$. Therefore the evidence suggests that in these salt marshes, the species' up-gradient zone boundaries were at random and their down-gradient zone boundaries overlapped regularly like tiles on a roof. The latter result is significant at the 0.2% level (two-tailed).

Table 6.1 The observed and expected median numbers of "empty" quadrats, lacking "last encounters" and hence not containing the estimated locations of down-gradient zone boundaries, in transects across a salt marsh. Shown here are results from 5 of 40 transects (Pielou and Routledge, unpublished data)[a]

		Number of "empty" quadrats		
Number of quadrats in transect Q	Number of spp in transect s	x	Expected median $M(x)$	Result $x < M(x)$
21	13	12	12.5	yes
32	22	20	18.5	no
23	14	12	13.5	yes
34	24	16	19.5	yes
36	23	18	21.5	yes

[a] For each transect, its last quadrat is not included among the Q recorded; and the last quadrat's species, which are among the s in the whole transect, are not treated as encountered for the last time within the transect; that is, the boundaries of these species are assumed to be uncut by the transect.

Now consider how the data may be used to decide whether the zones of the several species are mutually adjusted (as in Figure 6.6*d*) or independent (as in Figure 6.6*e*). A possible test is to classify each of the Q quadrats in a transect so that the data *from each transect* can be tabulated in a 2×2 table thus:

		Quadrat contains an up-gradient zone boundary		
		Yes	No	
Quadrat contains a down-gradient zone boundary	Yes	a	b	$a+b$
	No	c	d	$c+d$
		$a+c$	$b+d$	Q

Under the null hypothesis that the locations of up- and down-gradient boundaries are independent, the distribution of the frequency in any cell of the table is hypergeometric; hence the probability that the yes-yes frequency is a is

$$p_a = \frac{\binom{a+b}{a}\binom{c+d}{c}}{\binom{Q}{a+c}}$$

with $\max(0, |a-d|) \le a \le \min(a+b, a+c)$.
Evaluating and summing these probabilities for a given transect, we may determine $M(a)$, the expected median frequency in the yes-yes cell. Thus, putting $k = \max(0, |a-d|)$,
if

$$\sum_{a=k}^{M} p_a = 0.5, \qquad \text{then} \qquad M(a) = M$$

and if

$$\sum_{a=k}^{M} p_a < 0.5 < \sum_{a=k}^{M+1} p_a, \qquad \text{then} \qquad M(a) = M + \frac{1}{2}.$$

Some results—again a subset of the results from 40 transects across two Nova Scotia salt marshes—are shown in Table 6.2. It was found that $a > M(a)$ in 30 of the 40 transects. The evidence that competitive exclusion had caused mutual adjustment of the zone boundaries is therefore very strong. This conclusion does not conflict with the earlier conclusion—that up- and down-gradient zone boundaries had nonmatching patterns—since

Table 6.2 The observed and expected median numbers of quadrats containing both up- and down-gradient zone boundaries in transects across a salt marsh. Symbols for cell frequencies are as in the 2×2 table in the text. Shown here are results from 5 of 40 transects (Pielou and Routledge, unpublished data).[a]

Q	$a+b$	$a+c$	a	$M(a)$	$a>M(a)$
20	7	8	4	2.5	yes
31	13	11	6	4.5	yes
22	13	11	8	6.0	yes
33	13	17	6	6.5	no
35	18	13	8	6.5	yes

[a] The first and last quadrats in each transect were disregarded.

observations significantly at odds with the two null hypotheses could have been predominantly at different levels of the gradient.

This last point suggests that much might be learned from comparative studies of two quite different kinds of zoned vegetation. Thus there is zoned vegetation where plant cover of any kind grades into a surface where rooted vascular plants cannot grow, for example, deep fresh or salt water, permanent snow or ice, or true desert. And there is often zoned vegetation where a broad ecotone lies between one kind of vegetation and another. In both kinds of community the sorting effect of environmental gradients is operative. It seems likely that zoned communities will provide a better testing ground for many theories of community structure than will communities in uniform environments.

Chapter 7

Determinants of Diversity: Local Factors

7.1 Introduction

In the two concluding chapters of this book we consider some of the current theories on species diversity and how they are interrelated. "Diversity" will be equated with "number of species." Probably all community ecologists have mulled over the problem: why does this particular s-species community not have more, or fewer, species? Why should the number be s, will it change, and if so in which direction and why? Thus worded the questions do not define which of two separate worlds of enquiry they come from, and which of two separate families of possible answers is to be broached. In searching for causes of the fact that a community has a given number of species, we may look for a proximate or an ultimate cause. Both are legitimate and interesting fields of enquiry but it is important to keep the distinction between them clear.

In the context of some one "true" community at a particular time, it is reasonable to seek for proximate causes: that is, to ask how the s species-populations comprising the community manage to maintain themselves simultaneously on the limited resources available. By a "true" community is meant one whose member individuals interact, either directly or through a chain of other individuals, in a way that affects their individual lifetimes and chances of reproduction and survival.

In the context of areas of "geographic" extent, one may search for ultimate causes: for example, one may ask whether the turnover of species (in some large taxon such as an order or class) resulting from evolution and extinction is fast or slow; whether the rates are in balance or whether the number of species in the world exhibits a long-term trend; and why there should be a latitudinal gradient in species diversity.

Roughly speaking, the causes of small-scale short-term phenomena in the biosphere are the subject matter of ecology; and the causes of large-scale long-term phenomena are the subject matter of biogeography. One could say that ecology deals with proximate, and biogeography with ultimate, causes of community phenomena. Since all such phenomena are necessarily affected by causes at both levels, any student of community biology is concerned with both fields and alternates between them spontaneously. In discussing the factors that determine community diversity it is therefore inevitable, and indeed desirable, that the focus of interest should switch back and forth; but to prevent fruitless arguments in which the debaters are talking at cross purposes it is important to be clear at all times which level of causation the discussion is focused on at the moment. Insofar as the discourse can be restricted to one level at a time, Chapters 7 and 8 will review theories having to do with proximate and ultimate causes of diversity respectively, but artificial limits will be avoided. Some topics, such as the theory of the niche, are basic to both.

7.2 The Niche and Species-Packing

In a discussion of the theory of the niche, and the competitive exclusion principle, it helps to start with a list of what appear to be four incontrovertible statements (perhaps they could be called axioms) that all biologists would concur with. They are:

1. Every organism has a *fundamental niche*, defined as the set of all environmental conditions that permit it to exist. If the values of n factors (not necessarily independent) are relevant to the organism's survival, its niche can be represented (Hutchinson, 1958) as a hypervolume in an n-dimensional coordinate frame each axis of which corresponds with a

factor. (The truth of this assertion seems undeniable; whether it is useful is, of course, quite a different matter.)

2. No intrabreeding population of organisms (a *gamodeme*) having a single fundamental niche can continue indefinitely to grow in numbers. If environmental vicissitudes do not keep it in check, then intrapopulation competition will. (This statement says nothing about the relative importance of density-dependent and density-independent population regulation in nature; it states only that one or the other must take place.)

3. If two or more gamodemes with identical fundamental niches live continuously mingled in the same enclosed uniform area, and their numbers are regulated by shortages of the same limiting resources, all but one of the gamodemes will eventually become locally extinct unless changes in environmental conditions, or in the characteristics of the gamodemes themselves, supervene in time to prevent it. [This is the *competitive exclusion principle*, sufficiently hedged with conditions (I believe) to make it irrefutable.]

4. It follows that occurrence of several "apparently similar" gamodemes in a community must imply one or more of eight things:

(a) That despite appearances, the gamodemes' fundamental niches are not identical and each has a large enough *realized niche* (defined by Hutchinson (1965) as a set of conditions necessary for survival that it shares with no other gamodeme) to enable it to persist.

(b) That the community has not reached equilibrium.

(c) That environmental conditions are changing with time.

(d) That the gamodemes are changing with time.

(e) That the gamodemes are not resource-limited.

(f) That the space is nonuniform.

(g) That the space, whether uniform or not, is divided into subspaces by barriers (they need not be total) that prevent complete mingling, at all times, of the competing populations.

(h) That immigration into the area is possible.

Each of these eight ways of explaining cases of coexistence that seem in need of explaining can, and often do, lead to circular arguments in spite of Hutchinson's (1965) insistence that a niche be defined as "that thing in which two sympatric species do not live." In Sections 7.3, 7.4 and 7.5 we shall consider the ways in which mechanisms (b) through (h) permit competing gamodemes to coexist. The present section discusses mechanism (a); this entails consideration of the taxonomic nature of the population unit (called a gamodeme above) that is the "proprietor" of a niche, and what is meant by the "size of a niche."

First, as to the proprietor of a niche: it has been compellingly argued by Ehrlich and Raven (1969) that the groups of organisms on which natural selection operates are far smaller than taxonomic species. The true evolutionary units are local intrabreeding populations that share a common gene pool, and exchange of genes between these units is infrequent. The term *gamodeme* (Briggs and Walters, 1969), more commonly used by students of plant than of animal evolution, is the most useful name for them. Since these are the units that have distinct gene pools, they are also the units that "own" a common fundamental niche. That is, the proprietor of a niche is not a species in the ordinary taxonomic sense, but a gamodeme. A single species usually comprises many gamodemes and hence has many niches. There is no need for different gamodemes of one species to be geographically separated; several can occur in one area and belong to the same community. Examples have been reviewed by Briggs and Walters (1969), and by Ehrlich and Raven (1969) who describe numerically small, closely spaced gamodemes in, for example, butterflies, mice, and grasses.

Therefore, strictly speaking, gamodemes rather than species are the entities that should be considered when a community's diversity is to be measured. Precise measurements of diversity (taking account of the relative abundances of the "species"), or rough measurements (a simple count of the "species"), should treat gamodemes rather than taxonomic species as the units of which a community is composed. This is, of course, an unattainable ideal but its implications are worth contemplating. Separate gamodemes, which are (as a rule) noninterbreeding, may differ from one another hardly at all in visible, morphological characteristics, so that what is really a highly diverse community may be judged, from a list of its taxonomic species, to have low diversity. Two communities with conspicuously different apparent diversities could, conceivably, have identical "true" diversities (in the sense that each contained the same number of gamodemes in the same relative proportions) but manifest this diversity at different hierarchical levels. For example, using letters to differentiate species and subscripts to differentiate the gamodemes within them, two communities symbolized by

$$A \quad B \quad C \quad D \quad E \quad F \quad G \quad H \quad I \quad J \quad K \quad L$$

and

$$A_1\,A_2\,A_3 \qquad B_1\,B_2\,B_3 \qquad C_1\,C_2\,C_3 \qquad D_1\,D_2\,D_3$$

could be thought of as having the same total diversity manifested at different levels in the two cases.

Turning to the competitive exclusion principle, we have also to consider whether the entities that exclude one another are taxonomic species or gamodemes. If two truly intersterile species compete, and one excludes the other, competitive exclusion can certainly be said to have taken place. But separate gamodemes of a single species may be moderately interfertile (either physiologically or behaviorally); then outbreeding by the members of a gamodeme is not impossible, only comparatively unlikely, and it would presumably become less unlikely for members of a gamodeme that, because of competition, was becoming sparse. Thus a gamodeme in process of exclusion, instead of becoming locally extinct, might be merely submerged in an "if-you-can't-lick-'em-join-'em" process. The losing gamodeme would then make at least a small genetic contribution to the surviving population, and whether the loser should be described as wholly excluded is a matter of opinion.

We now consider the species-packing problem, that is, the question: how many species (or gamodemes, but we shall not use this term in what follows) can coexist in equilibrium? Other ways of phrasing the problem are: what limit is there to the similarity of coexisting species (MacArthur and Levins, 1967)? Or, how small can a realized niche be for its proprietor population to maintain itself?

The last version—in terms of niche volumes—has led to some diverting assertions, for example "wide niches will usually contain a greater total amount of food than narrow niches" (this comes verbatim from a published paper) which leads one to wonder whether the container envisaged is a niche or a freezer. A species' niche, in the current Hutchinsonian usage, is an n-dimensional hypervolume in phase space. Some, but not necessarily all (Wangersky, 1972) of its points can be mapped into the real space where a community lives. However, *all* the points in the community's real space can be mapped into the phase space and the intersection of this hypervolume with the niche hypervolume of a given species represents the usable niche of the species in the community concerned. Thus although the magnitudes of niche hypervolume and the corresponding occupied real volume are probably correlated to some degree, the correlation may be weak. (These points are further elaborated in Section 8.3).

This makes a direct attack on the species-packing problem very difficult; the attempts that have so far been made, by means of idealized mathematical models, do not appear to have led to useful insights. A promising alternative to direct attack is to investigate the relationship between the morphological variability of a species-population and the number of competing species it coexists with. Much work has been done on these lines; birds, especially, have been studied in this connexion since the length, width, and depth of the bill are easily measured and data can be obtained in large quantities by using museum specimens; also, bill dimensions are obviously closely related to feeding habits, and high variability in a population's bill morphology can reasonably be taken to imply that the population as a whole (not necessarily the individual birds) eats a varied diet. The variability may consist in continuous variation, discontinuous variation (polymorphism), or sexual dimorphism. From a review by Rothstein (1973) of these studies on bird morphology, there appears to be evidence showing that there is often significant negative correlation between morphological variability and number of sympatric competitors.

Investigations of *character displacement* have provided additional examples. Character displacement is the phenomenon in which members of two closely related species differ from each other markedly in areas where they are sympatric, whereas both converge towards a common form in areas where one occurs without the other. It is usually found (Mayr, 1973) that morphological variability is much lower in "divergent" species-populations coexisting in one area than in "convergent" populations of the two species living separately.

Variability within an intrabreeding species-population, that is, within one gamodeme, thus exemplifies diversity at a still lower hierarchical level than does between-gamodeme variability. In comparative studies of community diversity attention should be paid not only to differences between the species-diversities of communities and their possible causes, but also to the following problem, which is no less interesting: *what are the factors that determine the hierarchical level at which community diversity manifests itself and why should the level differ in different communities*?

7.3 The Coexistence of Competitors

Competitive exclusion is expected to occur when competing species-populations are too similar to coexist. Until the ecological similarity between species can be measured in a way that does not take competition into account, this definition will remain circular. All the same, it is interesting to explore mechanisms that appear to permit competitors to coexist even when we feel subjectively that one species "ought" to exclude the others. These mechanisms were listed in Section 7.2; in this and the following sections they are given more detailed treatment.

The first and most obvious possible cause for seemingly anomalous coexistence is that exclusion is occurring but is not yet complete. It can sometimes take a very long time. There are two different modes of competition among animals: *exploitation* and *interference*. Exploitation occurs when the limiting resource, usually food, is constantly accessible to all the competitors; this form of competition is common among small-bodied organisms with short generation times, typically invertebrates. Interference is competition in which the competing individuals occupy and defend territories, thus ensuring a supply of their needs for some time into the future; it is common among large-bodied organisms with long generation times, typically vertebrates; many species of mammals and birds are territorial, for example. Miller (1969) has argued that competitive exclusion occurs much more quickly, in terms of numbers of generations, when interference rather than exploitation is taking place. He has assembled results of eight laboratory investigations of two-species competition (which, perforce, used organisms that compete through exploitation) in which the exclusion of the losing competitor never occurred in fewer than 32 generations. Thus, prolonged coexistence of invertebrate competitors should not be thought surprising even when they occupy a habitat that is enclosed, spatially uniform, and temporally constant, still less when they live under natural conditions.

Environmental conditions never are constant for long, of course, except possibly in the deep sea, in caves, and in continuously wet rain forest. The effect on the competitive exclusion process of an alternating succession of dormant seasons and growing seasons depends on whether the population sizes of the species concerned grow large enough to become resource-limited before their growth is checked by deteriorating conditions. Theoretically, the alternating seasons would not prevent competitive exclusion if, each year, the sizes attained by the competing populations were regulated by the available amount of their common limiting resource. Then changes in the relative proportions of the competitors would be carried forward from one year to the next and exclusion would proceed, though with interruptions. However, if the growing season is too short, relative to the competitors' growth rates, for a season's populations to become large enough to be resource-limited, then cumulative changes, leading to competitive exclusion, will not occur. This mechanism may well explain the observed coexistence of several congeneric aphid species on a single species of host plant (Pielou, 1974*c*), for example; and the enormous number of coexisting species of phytophagous insects that live in birch bracket fungi (Pielou and Verma, 1968).

If cyclical variation in environmental conditions can permit competing species to coexist, so also can cyclical variation in the competing populations themselves. Pimentel et al. (1965) postulated the following process to account for some cases of successful coexistence: in a resource-limited two-species system, every individual is competing simultaneously with members of its own species and of the other species; in the larger population, natural selection will favor those individuals that are adapted to succeed at within-species rather than between-species competition; in the smaller population the opposite will be true. As a result the larger population will become smaller, and the smaller larger. The populations will seesaw, changing genetically as they do so. Pimentel et al. demonstrated the process experimentally using populations of houseflies and blowflies as competitors.

Species with similar requirements can coexist successfully provided none of their populations is resource-limited; the species are then potential, rather than actual, competitors and will remain so for as long as their population sizes are regulated by some other factor than shortage of the shared resource. Predators are often suspected of keeping potential competitors too sparse to affect one another. Definite proof that this was the case has been given by Paine (1966) who found that the experimental removal of sea-stars from an area of rocky shore led to the exclusion of species of chitons, limpets, and algae by mussels, barnacles, and goose-necked barnacles whose populations exploded as soon as control by the predatory sea-stars ceased. And by Harper (1969) who described the exclusion by a few grass species of other grasses and several species of dicotyledons that occurred in British chalk grasslands when rabbit grazing stopped suddenly; (the collapse of the rabbit populations was due to a myxomatosis epidemic).

The competitive exclusion process is sometimes prevented from going all the way to completion by the fact that sparsity confers an advantage on a dwindling population and saves it from dwindling further. For example, if a predator preys upon two or more mutually competitive prey species and, at any time, concentrates its attacks on the momentarily most abundant prey, the effect will be to maintain the prey populations in fairly equal proportions. This mechanism also operates at the subspecific level and hence maintains subspecific diversity. The process is known as *apostatic selection.* An example has been described by Clarke (1969): the two morphs of a dimorphic snail species differ from each other in the color and pattern of their shells; they are preyed upon by thrushes which tend to eat whichever is, at the moment, the commoner morph. The result is that each morph is left alone while the other is more abundant.

Paulson (1973) has argued that an analogous process can maintain polymorphism in predators and possibly accounts for the several color morphs that often occur in a single species in diurnal birds of prey; if the small mammals that the birds eat learn to recognize and avoid the commonest morph of their predator, one of the less common morphs will fare better until it, in turn, becomes familiar to the prey.

None of the several mechanisms described above, that can prevent competitive exclusion from occurring and that tend, consequently, to maintain high diversity within groups of ecologically similar species at the same trophic level, assumes that the habitat of the community concerned is other than uniform. As soon as we relax this restriction, the likelihood that similar species will coexist increases (it seems) by an order of magnitude at least. Nonuniform habitats can affect the communities that occupy them in a multitude of ways and Sections 7.4 and 7.5 discuss some of them.

7.4 Diversity in Heterogeneous Habitats

If the habitat of a community, instead of being strictly uniform, is made up of a mosaic (two- or three-dimensional) of different microhabitats or substrates, the chance that many similar species will be able to coexist is greatly increased. Competition among organisms whose "home ranges" are of the same order of size as the substrate patches or smaller will occur only among species sharing the same substrate; occupants of different substrates will rarely encounter one another and will not compete. For brevity, this may be called the *substrate-patchiness effect.* Small organisms (small, that is, relative to the substrate patches) with larger home ranges will also benefit from what may be called the *nook-and-cranny effect,* which will be discussed later in this section.

Larger organisms, whose home ranges embrace several or many substrate patches, may not be affected directly by the patchiness but, insofar as the smaller organisms on which they feed are more diverse, the larger organisms can develop specialized diets and hence avoid direct competition. This effect can filter up through several trophic levels. Further, patchiness itself is often, perhaps always, hierarchical. For example, in a cyclical vegetation mosaic (cf Section 5.2), each phase constitutes a microhabitat for some species of animals, say carnivorous insects; and each plant species within a phase (if there are more than one) and then each plant organ (roots, stems, leaves, fruits, and so on) within a species, constitutes a micro-microhabitat for smaller organisms, among them phytophagous insects that are the prey of the carnivores.

Examples of the substrate-patchiness effect abound. One has already been mentioned (Section 5.2), that of the different wood warbler species occupying a mosaic of habitats in spruce forests (MacArthur, 1958). Two more examples will be given. O'Neill (1967) described a remarkable case in which each of seven species of dilopods (millipedes), living in a small area of deciduous forest floor in leaf litter and fallen logs, had its own specialized microhabitat; these were: heartwood of logs; superficial wood of logs; between wood and bark; on underside of logs; on ground beneath logs; within leaf litter; on ground beneath litter. Abele (1974), studying seashore communities of decapod crustaceans (crabs, lobsters, crayfishes, and shrimps) found strong positive correlation between species number and the number of different substrate types available; thus the species were fewest on sandy beaches, then progressively more numerous in, respectively, Spartina marshes, mangrove swamps, coral communities, and the intertidal communities of rocky shores.

The nook-and-cranny effect occurs when the patches in a patchy habitat differ in the protection they afford to prey species being hunted by predators. Then, as Smith (1972) has argued, those habitat patches where prey (possibly of several species) can find refuge from predators, serve as sites where the prey populations can maintain themselves in spite of predation pressure. The only individuals exposed to predation are a "surplus" forced out of the refugia into less sheltered habitats by competition. The prey populations in the refugia steadily grow and "overflow"; and the unprotected overflow, which is surplus to the species' sustainable population level, is as steadily skimmed off by predators who themselves congregate in the patches where hunting is easiest. The difference between the substrate-patchiness effect and the nook-and-cranny effect is that in the former, different species are segregated into different habitat patches, whereas in the latter, several species share, and thrive in, the same kind of habitat patch. The nook-and-cranny effect can obviously manifest itself at a wide range of scales; the predators and prey could be hawks and small rodents, preying mantises and small insects, or carnivorous fishes and their diet of smaller fishes and invertebrates.

The importance of the nook-and-cranny effect in maintaining community diversity may be very great, as Smith (1972) has argued. When attempts are made to simulate community behavior with computer models that take account of species interactions, stability is difficult to achieve, the more so the more species are included. And a carefully adjusted stable model of Lotka-Volterra type will usually "collapse" (several species will go extinct) as additional adjustments are made, for the sake of realism, by allowing for stochastic fluctuations and delayed responses. Collapse can only be staved off by allowing, as well, for spatial heterogeneity and for the existence of "good" patches and "poor" patches with the latter being replenished as needed by migrants from the former. Not to allow for this effect is so unrealistic that it seems futile to pile up refinements on a uniform-habitat model. Many of the refinements will probably be of trivial importance vis-à-vis the overriding influence of spatial heterogeneity. Allowing for it is not simple however. As will be shown in Section 7.5, models of processes in patchy environments can yield unforeseen results.

7.5 Two Models Assuming Spatial Heterogeneity

We shall now consider two models of population interactions in patchy habitats. The first, due to Skellam (1951), shows that if the only sites where two competing species can live within an area are separate "islands" of a single kind, then even though the competitively stronger species may invariably exclude the weaker from an island invaded by both, both can nevertheless coexist indefinitely in the area as a whole provided the weaker species is sufficiently fertile.

The second model, due to Horn and MacArthur (1972) arrives at the same conclusion by an altogether different route. But it shows also, that if there are two different kinds of islands in an area, conditions may make it impossible for two species to coexist even though each species can exclude the other from one kind of island. (Throughout this section the word "island" will be used to connote an occupiable habitat patch surrounded by ground wholly inhospitable to both species.) The models are too simple to be realistic but this does not affect their usefulness in showing, at least qualitatively, what *can* happen in natural situations, and hence performing what is perhaps a model's only useful function in the context of many-species communities living in natural conditions (cf Section 6.2).

Skellam's (1951) model is as follows. Imagine two competing species of sessile organisms (for example, annual plants) that reproduce once a year and have life spans of a year or less so that there is no overlap of generations. Call the species A and B and suppose they occur in a region where they can establish themselves only on identical isolated islands (in the sense defined above). These islands are all of the same area; within each of them any number of disseminules (seeds or eggs) of both species can begin growth, but only one individual (of either species) can survive to reproductive maturity. Let species A be the stronger competitor and assume that whenever individuals of both species begin life in a single island, A invariably wins. Species B can therefore reach maturity only in islands in which it is the sole colonizing species (see Figure 7.1).

Now consider species A. Assume the region contains N islands altogether; that the average mature A-individual produces enough disseminules for there to be a mean of F per island from each of them; and that, at equilibrium, the expected proportion of islands that will be found to contain a single A-individual at the end of the growing season is Q. Then in the whole region, the mean annual input per island of species A's disseminules will be NQF and assuming they are dispersed at random, the expected number falling in each island will be a Poisson variate with parameter NQF. Therefore the expected number of islands receiving at

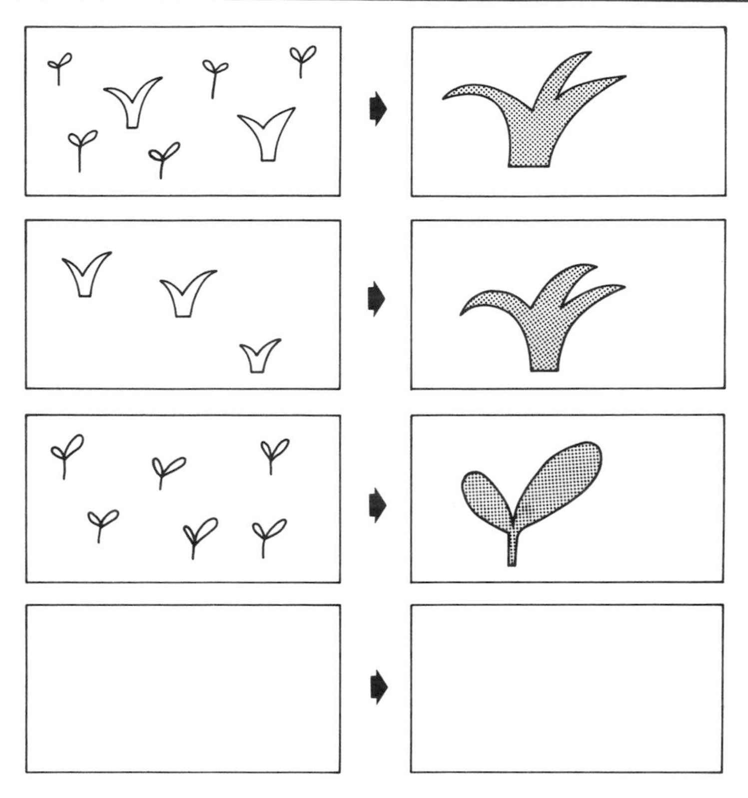

Figure 7.1 Each square represents an "island." The four possible species combinations in spring are shown on the left; how they will appear in the fall, on the right.

least one A-disseminule, and hence destined to contain exactly one mature A-individual at the end of the growing season, is $N(1-e^{-NQF})$. But this, by definition, is NQ. Hence we have the identity

$$NQ = N(1-e^{-NQF})$$

whence

$$F = \frac{-1}{NQ} \ln(1-Q). \tag{7.1}$$

The foregoing is also true when species B is present; since A always defeats B in competition, B's presence makes no difference to A's success.

Now consider how species B will fare. Since NQ of the N islands are expected to be preempted by A-individuals, only $N(1-Q)$ will remain available for B's. Let the expected proportion of all the islands that will be occupied by mature B's at the end of the season be q; and let the mean number of disseminules produced per individual per island be f; (thus the parameters q and f for species B are analogous to Q and F for species A). Then

$$Nq = N(1-Q)(1-e^{-Nqf})$$

whence

$$f = \frac{-1}{Nq} \ln \frac{1-Q-q}{1-Q}. \tag{7.2}$$

For species B to escape competitive exclusion, f/F must be great enough to ensure $q>0$. We may find the minimum fertility necessary to ensure B's continued coexistence with A (as a multiple of A's fertility) by determining $\lim_{q\to 0} f/F$. From (7.1) and (7.2) it will be found (using l'Hospital's rule) that

$$\lim_{q\to 0} \frac{f}{F} = \frac{-Q}{(1-Q)\ln(1-Q)}. \tag{7.3}$$

Thus the condition that B shall maintain itself by occupying, each year, islands that happen not to have received any of A's disseminules is that f/F shall exceed the term on the right in (7.3). Provided this condition is met, competitive exclusion will not occur.

The Horn and MacArthur (1972) model starts by postulating two types of "island" (again using the word as defined previously), types 1 and 2.

Let the two species, called P and Q, occupy proportions p_i and q_i respectively of the type i islands for $i=1, 2$. Both species emigrate from occupied to unoccupied islands of both types and so colonize those in which they were not already present. And both species sometimes suffer local extinctions (both spontaneous and because of competitive exclusion) in islands which they have occupied. Thus in any one island each species comes and goes.

Now consider the rate of change of q_1, the proportion of the type 1 islands where Q occurs. In what follows the m's, e's and c's are constants of proportionality pertaining to migration rates, extinction rates, and competitive success rates respectively. It is assumed that the rate of migration of a species, say Q, *from* type i islands depends on the proportion occupied, namely q_i; and that the rate of migration of Q *into* type i islands where it is not already present depends on the proportion unoccupied, namely $(1-q_i)$.

Therefore the effect of migrations will be to increase q_1 at a rate of $m_{11}q_1(1-q_1)+m_{21}q_2(1-q_1)$. Here m_{ij} is the constant of proportionality applying to migration from type i to type j islands.

At the same time Q is assumed to become extinct spontaneously in type 1 islands at a rate e_1q_1.

Further, assume the two species invade islands independently, so that the number of type 1 islands in which they co-occur is p_1q_1; assume also that in a proportion c_1 of these jointly occupied islands, species P is competitively stronger and excludes species Q. This causes q_1 to decrease at a rate of $c_1p_1q_1$.

Summing the effects of migration, spontaneous extinction, and competitive exclusion, we see that

$$\frac{dq_1}{dt}=(m_{11}-e_1-c_1p_1)q_1+m_{21}q_2-m_{11}q_1^2-m_{21}q_1q_2.$$

Next, write

$$p_i=\hat{p}_i \qquad (i=1, 2) \qquad \text{when} \qquad q_1=q_2=0.$$

That is, the $\hat{p}$ values are the proportions of the two types of islands occupied by species P at equilibrium in the absence of species Q. If Q were now to invade, the initial invasion rates, with q_1, $q_2 \simeq 0$, and terms in q_1^2 and q_1q_2 negligible, would therefore be

$$\frac{dq_1}{dt}=(m_{11}-e_1-c_1\hat{p}_1)q_1+m_{21}q_2 \tag{7.4}$$

and analogously

$$\frac{dq_2}{dt}=(m_{22}-e_2-c_2\hat{p}_2)q_2+m_{12}q_1.$$

Thus a *sufficient* condition for Q to invade the whole region, which requires $dq_i/dt>0$ for $i=1$ or 2 is

$$m_{11}>e_1+c_1\hat{p}_1 \qquad \text{or} \qquad m_{22}>e_2+c_2\hat{p}_2. \tag{7.5}$$

Notice that if all the islands are of type 1 and if, also, $c_1=1$ (so that P always excludes Q from jointly occupied islands) then, provided

$$m_{11}>e_1+\hat{p}_1 \tag{7.6}$$

Q can invade the area in spite of its competitive inferiority in all the islands. This result is qualitatively the same as Skellam's. The same possibility has also been demonstrated by Levins and Culver (1971).

We now demonstrate the other consequence of this model, namely that Q can fail to invade the area even when there are two types of islands and Q is competitively superior to P in type 2. (That is, Q wins only Pyrrhic victories over P).

First consider the isoclines $dq_i/dt = 0$ for $i = 1, 2$. From (7.4) it is seen that they are

$$q_2 = \frac{q_1(e_1 + c_1\hat{p}_1 - m_{11})}{m_{21}} \qquad \text{when} \qquad \frac{dq_1}{dt} = 0$$

and

$$q_2 = \frac{q_1 m_{12}}{e_2 + c_2\hat{p}_2 - m_{22}} \qquad \text{when} \qquad \frac{dq_2}{dt} = 0.$$

Plotting these isoclines in a phase-space diagram as in Figure 7.2, it is seen that q_1 or q_2 can increase continuously (that is, species Q can invade the area) only if the slope of $dq_2/dt = 0$ exceeds that of $dq_1/dt = 0$; that is, if

$$\frac{m_{12}}{e_2 + c_2\hat{p}_2 - m_{22}} > \frac{e_1 + c_1\hat{p}_1 - m_{11}}{m_{21}}. \tag{7.7}$$

This is the *necessary* condition for invasion of the region by Q. It may hold even if neither condition in (7.5) is met.

Now notice that

$$\frac{m_{22}}{m_{21}} = \frac{m_{12}}{m_{11}}. \tag{7.8}$$

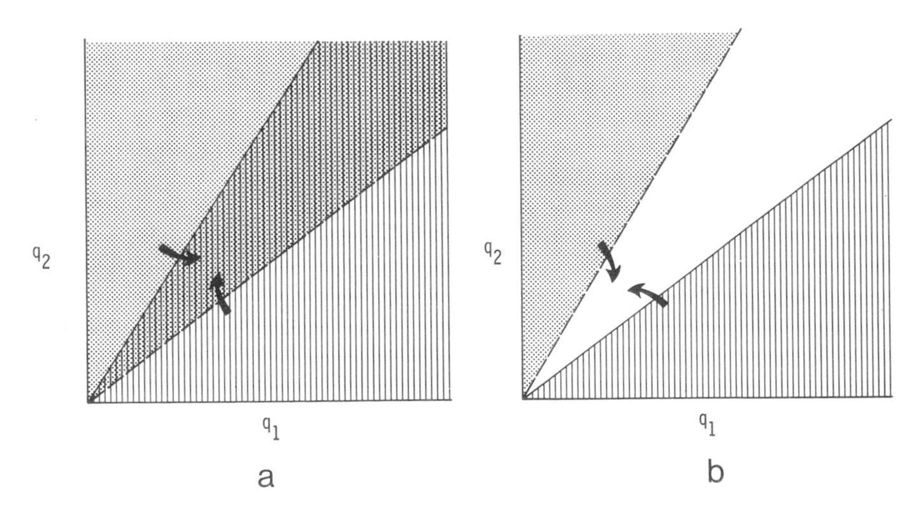

Figure 7.2 The (q_1, q_2) phase space. In the stippled region above the $dq_1/dt = 0$ isocline (dashed line) q_1 increases; in the hatched region to the right of the $dq_2/dt = 0$ isocline (solid line) q_2 increases. The arrows are population trajectories. The isoclines are assumed to be linear only in the neighborhood of $(q_1, q_2) = (0, 0)$; thus the graph shows only a small part of the phase space close to the origin. (Redrawn from Horn and MacArthur, 1972.)

This follows since both ratios are equivalent to the ratio:

$$\frac{\text{chance of colonizing a type 2 island}}{\text{chance of colonizing a type 1 island}}$$

which is the same for both species.

Using (7.8) to eliminate m_{12} and m_{21} from (7.7), the latter inequality becomes

$$m_{11}(e_2+c_2\hat{p}_2)+m_{22}(e_1+c_1\hat{p}_1)>(e_1+c_1\hat{p}_1)(e_2+c_2\hat{p}_2). \tag{7.9}$$

If we now assume that P always defeats Q in type 1 islands, and that Q always defeats P in type 2 islands (that is, $c_1=1$ and $c_2=0$) (7.9) becomes

$$m_{11}e_2+m_{22}(e_1+\hat{p}_1)>(e_1+\hat{p}_1)e_2 \tag{7.10}$$

as the *necessary* condition which will permit Q to invade the region.

Reversing the inequality in (7.10) thus gives the condition in which Q will *not* be able to invade the region in spite of its competitive superiority to P in one of the habitat types.

The model thus permits the following two nonobvious conclusions to be drawn:

1. Two species *can* coexist indefinitely in a single habitat even when one invariably outcompetes the other, provided certain conditions are met.
2. An invading species may sometimes *fail* to establish itself in a region, even if there are habitat islands in which it is superior to the only competitor.

It follows that the competitive exclusion principle is an unreliable guide to community processes in natural, spatially heterogeneous habitats.

Chapter 8

Determinants of Diversity: Global Factors

8.1 Introduction

When the diversity of an ecological community is studied at close range the fact that usually excites interest and demands explanation is the successful (or, at any rate, prolonged) coexistence, in a small area, of several similar species. If it can be shown, as it usually can, that their requirements are not identical, or equivalently that each has a distinct realized niche, then there is no reason why they should *not* coexist. But at the same time, there is no immediately obvious reason why they should. Thus to explain why, in a given case, competitive exclusion is not occurring is not the same as explaining why coexistence *is* occurring. In attempting to explain the latter, we can no longer limit consideration to small areas and what may be called "local communities"; we must take account of larger regions, that is, regions of "geographical" extent, large enough to be regarded as the source from which a local community's species-complement is drawn. There is no clear boundary, of course, between "small areas" and "large regions"; linking together the separate bodies of theory that have been developed to explain diversity in the two contexts is now an active field of research which holds great promise; some of the advances so far made are discussed in Section 8.4. To begin with, however, in Sections 8.2 and 8.3, we focus attention on diversity in large regions and with geographical trends in diversity values.

There are several such trends, of which the most prominent is the well known latitudinal gradient in species diversity. Various aspects of the phenomenon are the topics of many discussions in the literature, for example Connell and Orias (1964), Simpson (1964), Pianka (1966), Stehli (1968), and numerous earlier papers listed in the bibliographies of these papers. In the majority of groups examined, species-richness increases, often dramatically, along a gradient from pole to equator. For example, the phenomenon is exhibited by numerous groups of plants and birds, by North American mammals (Simpson, 1964), littoral mosllusks (Abbott, 1968), and mosquitoes and butterflies (Stehli, 1968). Very rarely, and then only in rather narrowly defined groups, the gradient is in the opposite direction with species-richness increasing poleward within a latitude belt before decreasing again at still higher latitudes; for example this is true of penguins in the southern hemisphere and of the alcids (auks, razorbills, murres, guillemots, puffins) in the northern hemisphere (Stehli, 1968); and also, in the sublittoral and shallow shelf infauna (Thorson, 1957).

Another well-documented trend in the diversity of terrestrial species groups is that from high diversity in large continental landmasses to low diversity in small oceanic islands (MacArthur and Wilson, 1967). A somewhat less clear-cut manifestation of what is presumably the same effect is exhibited by the contrast in species richness shown by the mammals of the interior of North America and the peninsulas around the periphery of the continent (Simpson, 1964). This "peninsula effect" is also shown by the deciduous trees of Canada; there are 31 species in the Atlantic provinces and 50 species in a comparable area at the same latitude (north of 43.5° N) in Ontario (Hosie, 1969).

Among marine animals, in the bivalves and polychaetes of soft sediments, Sanders (1968) has demonstrated that diversity decreases from high values in tropical shallow seas, through successively lower values in the deep sea on the continental slope, in tropical estuaries, in boreal shallow seas, and in boreal estuaries. He has also shown (Sanders, 1969) a diversity contrast between boreal shallow water communities of these animals occurring in a continental climate (off the New England coast) where diversity is low, and in a maritime climate (off the state of Washington) where diversity is comparatively high.

Before embarking on detailed accounts of the theories that have been proposed to account for these geographical differences in species diversities, three points need to be mentioned. I shall not differentiate between so-called within-habitat diversity and between-habitat diversity as the distinction seems to obscure rather than clarify most discussions; what is a within-habitat environmental difference for large organisms is a between-habitat difference for small ones, and a clear dichotomy between them is obviously impossible.

Secondly, the notion that habitats exhibit greater structural complexity in low latitudes than in high and hence give more scope for the substrate-patchiness and nook-and-cranny effects (see Section 7.4) will also be given short shrift. There is no reason to suppose that any latitudinal gradient exists in the structural complexity of inorganic habitats; in living habitats, chiefly vegetation, it very probably does but this should be treated as a manifestation of the phenomenon whereby diversity in one group of organisms affects that in another.

Thirdly, an obviously important cause of greater within-region diversity in low latitudes than in high deserves mention here but will not be discussed further since it is wholly noncontroversial. This is the fact that a tropical mountainside offers a far greater range of environments than does a mountain in high latitudes. The contrast in conditions between summit and lowland for a given range of altitudes is obviously more pronounced the lower the latitude.

8.2 The Stability-Predictability-Productivity Family of Hypotheses

Arguments to show that high environmental stability, high environmental predictability, and high productivity are all conducive of high ecological diversity are probably familiar to most community ecologists. It seems unreasonable to treat these causes separately; they seem, rather, to constitute a suite of interdependent causes, as will be shown.

First, an attempt should be made to clear up the confusion that has arisen in the literature because of the use of the word "stability" without qualification. It can refer to *environmental stability* or *community stability*. There are strong reasons for believing that high environmental stability leads to high diversity, and subsequently we shall consider these reasons and how the term environmental stability is best defined.

The nature of the interdependence between diversity and *community stability*—or, equivalently, system stability; or the simultaneous population-stability of all a community's member populations—is best defined in terms of the behavior of the community when the absolute or relative sizes of its component populations are perturbed. Thus a community may be defined as unstable or stable according as a perturbation to the system tends to amplify itself or die out; [for a very full discussion, see May (1973)]. On the basis of subjective and qualitative arguments, it was for many years believed by the majority of ecologists that the more complex a community (that is, the more numerous its species and the more intricate their interrelationships) the greater the community's system-stability would be: for if each species could rely on many rather than few food sources, and be regulated by many rather than few predators, the eggs-in-one-basket effect would be minimized. As a result, high diversity would cause high community stability. This may indeed be so, but faith in the theory has wavered since May (1973) pointed out to ecologists that community stability is not a *mathematical* consequence of high species diversity. The contrary is true. To put the matter as briefly as possible: in most mathematical community models, made up of sets of differential equations, the larger the number of species the narrower the permissible ranges for the various parameters (interaction coefficients) if the system is to be stable; [see also Strobeck (1973) and Maynard Smith (1974)].

One (or both) of two conclusions follow. If, as seems to be the case, high diversity and high community stability are in fact positively correlated, then species interaction coefficients must tend, presumably because of natural selection, to lie within the restricted ranges necessary for the maintenance of community stability. The other (and, it seems to me, more likely) explanation for the correlation is this: high *environmental* stability leads to high community stability which, in turn, permits (but is not caused by) high diversity. The chain of cause and effect may be represented thus:

$$\text{Environmental stability} \rightarrow \text{Community stability} \rightleftarrows \text{High Diversity.}$$

The slashed arrow represents the possible tendency for high diversity to lessen community stability and hence to have an effect opposite to, but not as strong as, that of environmental stability. And even this weakening of environmental control will not happen if (1) the interaction coefficients have suitably selected values; or (2) spatial heterogeneity reduces the number of effective interactions among species. As was argued earlier (Section 7.4), many-species models that make no allowance for spatial heterogeneity are totally unrealistic. Such models, despite their unrealism, are often useful in suggesting interesting and unexpected phenomena that *can* occur, but not in persuading us that anything can*not* occur. Such prohibitions can always be made to disappear by modifying the assumptions. Consequently, *models reveal possibilities but not impossibilities.*

Now to consider the effects on diversity of environmental stability and predictability: debate on the definitions of terms is again possible here, of course. For the purpose of the present discussion* the following definitions will be used. If temporally regular (periodic) environmental fluctuations are of large amplitude, the environment will be called unstable; (and if of small amplitude, stable). Since by far the most important temporally varying factors of the environment are climatic, and since also the only important period in the present context is the year, we are in fact concerned with seasonal variability. Thus environmental instability and marked seasonality may be treated as synonymous; and since *seasonality* (with appropriate adjectives) is a less ambiguous word than *instability*, it will be used in what follows.

The colloquial meaning of *predictability* is perfectly clear. To be more precise, we shall say that an environment is more or less predictable according as the variances of the period and amplitude of environmental fluctuations are small or great. There is no need to separate amplitude variance from period variance for our present discussion, nor to labor the distinction between seasonality and unpredictability. From the point of view of their effects on diversity, there are two chief ways in which seasonality and unpredictability act to decrease it. These are: (1) To survive in a seasonal or unpredictable environment, a species needs to be "flexible in its responses" or, equivalently, to have a "wide niche"; as a result, a given type of habitat can contain fewer niches and hence fewer species the more strongly seasonal, or unpredictable, the conditions. Detailed discussions of this highly debatable statement are deferred to Section 8.3 and will not be mentioned further here. (2) Only those species that can quickly and surely recover from catastrophic die-backs can persist where such die-backs are likely to occur, namely in seasonal or unpredictable environments; thus species-populations that cannot build up again after they have been severely depleted will be automatically excluded from catastrophe-prone environments.

It has also been argued that the member species of low diversity communities in unpredictable environments will tend to be different in kind from those in high diversity communities in predictable environments. The former will have been selected for high fertility, to ensure quick recovery from population crashes following climatic catastrophes, and the

* I am not attempting to judge, or to choose among, other, more precise definitions; nor to propose methods for measuring environmental stability and predictability.

latter for qualities ensuring success in interspecific competition. These two kinds of species have been called, respectively, "*r*-selected" and "*K*-selected;" (see MacArthur and Wilson, 1967; and Pianka, 1970). Besides having a high value of *r*, the intrinsic rate of natural increase, *r*-selected species are expected to reach reproductive maturity sooner and to have shorter life spans and smaller body sizes than related *K*-selected species. It is also supposed that the population sizes of *r*-selected species vary widely but tend to be kept below their saturation levels by density-independent physical factors; whereas the population sizes of *K*-selected species are assumed to be much less variable and to be maintained at their saturation levels (*K*-values) by competition. Low diversity communities in which *r*-selected species predominate have been described as *physically controlled*, and high diversity communities in which *K*-selected species predominate as *biologically accommodated* (Sanders, 1968, 1969).

How widely applicable these generalizations are remains to be discovered. They were originally proposed in the context of terrestrial communities and may be inapplicable to some marine communities. Thus sometimes, as Thorson (1950) has argued, the prerequisite for success in high latitudes will not be the production of numerous small offspring, for these may not have time enough in a short growing season to assimilate the food necessary to permit their survival through the succeeding winter. A better chance for survival will, instead, belong to species that produce small numbers of large offspring or lay small numbers of large eggs; the young of such species will be less fragile and better able to survive when the cold season comes. As an example of a group of organisms adapted in this way, Thorson (1950) (and see Eltringham, 1971) describes the reproductive modes of marine gastropods belonging to benthic shelf communities. Some species lay numerous small eggs from which small larvae hatch, and these require a long period as pelagic planktotrophs before settling. In other species only small broods are produced, either ovoviviparously or in large, yolk-filled eggs, and the young need no pelagic period in order to complete their growth. Thorson found that the proportion of species of the latter kind (with no pelagic period in their life cycles) varied from 100% in communities off the coast of East Greenland to only 15% in waters off Bermuda. In these gastropods, therefore, *r*-selected species preponderate in low latitudes and *K*-selected species in high. Two of the qualities—high fertility and the ability to mature rapidly—that would presumably be advantageous for organisms living in high latitudes are, in fact, irreconcilable.

If differences in seasonality and unpredictability are the chief causes of diversity differences, then to account for the latitudinal diversity gradient it is necessary to show that these properties, or their effects, vary with latitude. Seasonality certainly varies with latitude in general, although exceptions are not uncommon; (for example, tropical climates with contrasting wet and dry seasons are strongly seasonal; maritime climates in middle latitudes are often fairly equable). Unpredictability is not obviously a characteristic of high latitudes, but its effects may be. Often it is the level of some one climatic factor of overriding importance (usually temperature) relative to a threshold level that governs the fate of a population; a crash occurs whenever the level falls below the threshhold. Moreover, temperature threshholds may be the same (or nearly so) for a large number of species. Thus even if identical patterns of temperature variation were to occur in two different latitudes, the frequency with which temperatures dipped below this threshhold would be much greater at the higher latitude where the mean temperature would be lower (see Figure 8.1). Identical degrees of unpredictability would therefore be more damaging at high latitudes; in other words, the intensity of *r*-selection would increase with latitude. (This does not, of course, explain why there are fewer *r*-selected species in high latitudes than there are *K*-selected species in low; we return to this point in Section 8.3.)

Two additional reasons for the fact that diversity is greatest in the tropics may be the following. A long warm season favors the presence of more trophic levels in a community than does a short one; there is always a delay between the buildup of a prey population and that of its predators

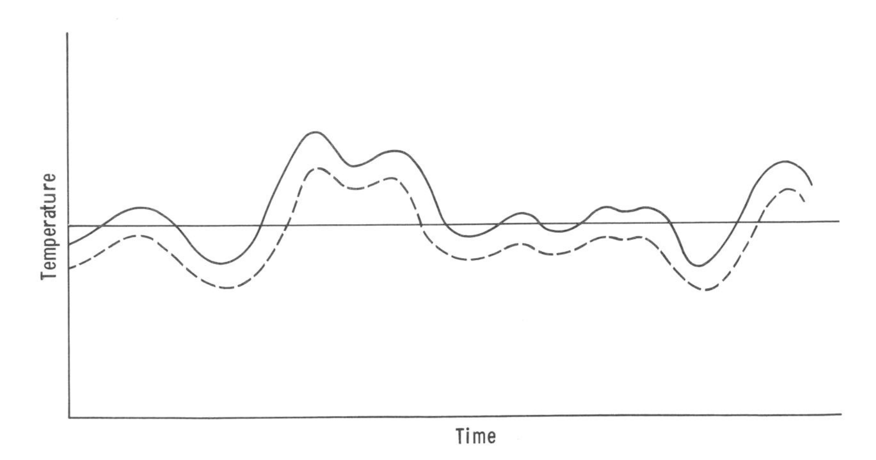

Figure 8.1 Identical patterns of temperature variation about a high mean value (solid line) and about a low mean value (broken line). The straight line represents a threshhold value common to many species.

and a long growing season thus gives time for populations at a greater number of trophic levels to complete their annual cycle of growth.

The second reason was suggested by Janzen (1967); it applies to the fauna and flora of mountainous regions in the tropics. He argues that in such country high mountain passes constitute an impassable barrier to the dispersal of all species not adapted to endure, at least temporarily, the relatively rigororous climatic conditions at the heads of the passes; most of the region's species can survive only in the continuous warmth of valley bottoms where they are never exposed to cold and thus have never been selected for the ability to withstand it. As a result, each of a species' intrabreeding populations is permanently trapped in a single valley, no mingling with other populations occurs, speciation proceeds rapidly and a region with many isolated valleys will be rich in species. In contrast, in a region of similar relief in a temperate climate, where winters are cold even in valley bottoms, all the species have been selected to survive low temperatures and therefore the high passes between valleys do not constitute much of a barrier to dispersal.

The argument that the high diversity of tropical communities is strongly influenced by the high productivity or, strictly speaking, the high level of solar energy input, as well as by weak seasonality, is due to Connell and Orias (1964). They suppose that diversity is directly proportional to the rate of energy flow; this determines the abundances of species-populations, and hence the sizes of their gene pools and their chances of subdividing into new species. The weak seasonality of wet tropical climates reinforces this effect. All the energy assimilated by living organisms is allocated to two purposes: either it is used for reproduction and growth, that is, for population expansion, in numbers or biomass or both; or it is used, and dissipated, in the maintenance of homeostatic regulatory processes. The less seasonal the climate, the smaller the fraction of assimilated energy that must be allocated to maintenance and hence the larger the fraction available for population growth; thus populations will quickly grow large and undergo speciation. An additional effect of the high rate of primary production is that animal populations can obtain all the food they require in home ranges of small area; their resultant low mobility reduces population mingling and hastens speciation.

The argument above assumes that high productivity on the one hand, and weak seasonality with high predictability on the other, tend to go together and that both contribute to high diversity as well as reinforcing each others' effects. In equatorial rain forest this may well be true; but for many communities the environment is highly predictable though unproductive (for instance in the deep sea and in caves) or else highly productive though unpredictable (for instance in tropical estuaries, oceanic upwellings or organically polluted waters). From the fact that diversity tends to be high in deep sea benthic communities (Sanders, 1968) and low in estuaries, upwellings (Sanders, 1969), and polluted waters (Williams, 1964) it seems reasonable to conclude that predictability is a more important factor than productivity in regulating diversity. It is true that high predictability cannot compensate for excessively low productivity as, for example, in caves (Poulson and Culver, 1969); where life is barely sustainable there is hardly any biomass to become diversified. Indeed one could say that caves, in spite of their environmental constancy (in those parts beyond the reach of floods) provide highly unpredictable environments in that only very minor fluctuations in the small amount of food available for cave animals will reduce the supply below the threshhold requisite for survival.

8.3 Variations in Niche Widths and Overlaps

Sections 8.1 and 8.2 ignored the following problem: in species-rich regions (such as the tropics), is the overall high diversity due to the fact that single communities are very diverse, or does it merely result from the presence in these regions of numerous separate communities each with a different complement of member species but each, by itself, only moderately diverse? In the wet tropics, without a doubt, both components of regional diversity are important; individual communities tend to have high diversity and, in addition, individual species tend to have small geographic ranges. Geographic ranges are not determined solely by climate of course; the shapes and arrangement of the continents are presumably as important as climate in determining them; we return to this point in Section 8.5.

The present section is concerned with the following question: why should there be regional differences in the mechanisms (discussed in Chapter 7) that control the diversities of single communities? In particular, does the great diversity of terrestrial communities in warm, equable climates imply that the numerous coexisting species have "narrow niches" (that is, are ecologically specialized); or is it that their niches, though tending to be of the same width as those in low diversity communities, can overlap one another without exclusion occurring?

Evidence that bird species tend to have narrow niches in the tropics seems fairly strong. Cain (1969) has argued that this is to be expected on common-sense grounds: the continually changing food supplies of a temperate climate force a bird, if it is not to starve, to switch from one diet to another several times in a year and only a "generalist" can do this. In the wet tropics, however, there is no advantage in being a generalist; the virtually constant environmental conditions ensure the perpetual availability of one type of food (soft fruit, for example) and no risk of starvation attaches to having a narrowly specialized diet. This argument cannot be extended, of course, to species (such as most insects) that become dormant in seasons when their usual food is unavailable.

The tremendous diversity of the tree species in wet tropical forests is more of a puzzle (Richards, 1969). Large numbers of closely related, morphologically similar, tree species are found mingled together in apparently uniform habitats and it is not clear why competition should not lead to the exclusion of at least some of them. The two mechanisms that seem most likely to account for high diversity in such plant communities are: (1) that habitats are far less uniform than they appear, at least so far as seeds are concerned; as Harper (1967) has shown, some plant species are exceedingly specific in the conditions they require for successful germination of their seeds; seemingly trivial microhabitat differences are decisive in determining where a plant of a particular species can establish itself and no trace of these minute differences remains detectable for long since it is the seedlings, not the mature plants, that are strongly niche-specific. And, (2) that "predation" by insects and pathogens keeps population sizes regulated so that inter-tree competition is much less fierce than it seems Janzen (1970).

There seems to be no reason, however, why these two arguments should apply with more force to tropical than to temperate forests. In any case, Richards (1969) doubts whether plant communities in general tend to be very much more diverse in the tropics than elsewhere. Although this is certainly the case with forest tree communities, it appears not to be true of communities such as sclerophyllous shrub vegetation or species-rich marshlands. The competitive exclusion principle has never been as appealing to plant as to animal ecologists.

It is clear that discussions based on qualitative observations, interesting though they are, will never lead to convincing conclusions as to whether it is "niche narrowness" or "niche overlap" that permits closely related species to coexist. Even when (as with many tropical birds) there is strong evidence to believe that their niches are narrower than those of their relatives in the temperate zones, we still do not know whether this is the sole reason for higher community diversities or whether niches tend to overlap to a greater extent as well.

It is therefore desirable to have an objective method for measuring niche width and overlap. It will be recalled that (in Section 7.2) we defined (following Hutchinson) a species' *fundamental niche* as the set of all environmental conditions that permit it to exist; and its *realized niche* as the subset of this set that it shares with no other species. The latter definition presupposes that competitive exclusion has occurred; it is more convenient to redefine a realized niche as the set of conditions that a particular species does in fact experience, regardless of whether other, competing species share all or part of the same niche. Thus "blurring" (the continual immigration of a species into a real habitat space where it is inferior to its competitors; see Section 6.2) is one of the mechanisms that may cause several species' realized niches to overlap.

Finally, allowance should be made for the fact that not every set of environmental conditions a species experiences (equivalently, not every point in the species' realized niche) is equally favorable to its survival and reproduction. Therefore, although "niche width," as it has come to be known, could be approximately quantified as the n-dimensional hypervolume of the realized niche, a more precise measure would be one that weighted each point in this hyperspace in a way that allowed for the varying degrees of success of the species in the community concerned. Thus niche width could be rigorously defined* as *a success measure on an environmental hyperspace.*

A point that sometimes causes confusion is easily cleared up. To speak of a species as having a niche does not imply that an individual of the species spends its whole life in the same habitat or under the same conditions. Allowance for changed habits and behavior at different stages in the life cycle is made by letting time be one of the n variables used to define each point in niche hyperspace. It is also sometimes argued that niche specification could theoretically be made so precise that niche overlap would be impossible merely in virtue of the fact that two solid bodies cannot simultaneously occupy the same physical space. This objection is easily countered by noting that corresponding to each point in niche hyperspace there is often a continuum of points in real space.

As the foregoing discussion shows, the niche is a highly abstract construct and to measure it is manifestly impossible. Attempts to do so seem futile. We can lower our sights, however, and, for suitably chosen organisms, make observations yielding quantitative measurements that can reasonably be interpreted as measures of niche width and overlap.

For example, consider a group of r, say, related species of sessile or sedentary organisms that occupy discrete, recognizable, easily classifiable habitats of c different kinds.† The average "range of occupied habitats" of

* To avoid possible ambiguity, this definition is deliberately differentiated from that of Levins (1968) who defined the "niche as a fitness measure on an environmental space." He appears to take the space of a fundamental niche as the space over which "fitness" is measured.

† These symbols are convenient as the data tabulation will be into r rows and c columns.

each member species of the group, relative to that of the group as a whole, can be measured and can be regarded as a measure of mean niche width (again, of each species relative to that of the whole group of species). The procedure for obtaining numerical results is formally as follows (Pielou, 1972). Suppose observations have been made of the number of occurrences of each species in each kind of habitat. An "occurrence" is the presence of one or more individuals of a species in a habitat; the numbers of individuals associated with each occurrence is disregarded. The observations are tabulated in an $r \times c$ *occurrence matrix* thus:

		Habitat kind					
		1	$\cdots$	j	$\cdots$	c	
	1	n_{11}	$\cdots$	n_{1j}	$\cdots$	n_{1c}	$n_{1.}$
	$\cdot$	$\cdot$		$\cdot$		$\cdot$	$\cdot$
	$\cdot$	$\cdot$		$\cdot$		$\cdot$	$\cdot$
	$\cdot$	$\cdot$		$\cdot$		$\cdot$	$\cdot$
Species	i	n_{i1}	$\cdots$	n_{ij}	$\cdots$	n_{ic}	$n_{i\cdot}$
	$\cdot$	$\cdot$		$\cdot$		$\cdot$	$\cdot$
	$\cdot$	$\cdot$		$\cdot$		$\cdot$	$\cdot$
	$\cdot$	$\cdot$		$\cdot$		$\cdot$	$\cdot$
	r	n_{r1}	$\cdots$	n_{rj}	$\cdots$	n_{rc}	$n_{r\cdot}$
		$n_{\cdot 1}$	$\cdots$	$n_{\cdot j}$	$\cdots$	$n_{\cdot c}$	N

Here

n_{ij} is the number of occurrences of species i in habitats of the jth kind.

$n_{i\cdot} = \sum_j n_{ij}$ is the number of occurrences of species i in habitats of all kinds.

$n_{\cdot j} = \sum_i n_{ij}$ is the number of occurrences of any of the group of species in habitats of the jth kind.

$N = \sum_i \sum_j n_{ij}$ is the total number of occurrences of all r species in all c kinds of habitat.

We can regard the N occurrences as a collection of N observations that can each be classified as belonging to one of rc classes. The diversity of these N observations can therefore be measured in the same way that the diversity of any collection of classifiable things can be measured.

Thus suppose, as a deversity measure we use Brillouin's index (see Section 1.4). Then the diversity of the N observations (equivalently, of the $r \times c$ occurrence matrix), which we shall denote by $H(AB)$ is

$$H(AB)=\frac{1}{N}\log\frac{N!}{\prod_i\prod_j n_{ij}!}$$

Now let

$$H(A)=\frac{1}{N}\log\frac{N!}{\prod_i n_{i\cdot}!}$$

$$H(B)=\frac{1}{N}\log\frac{N!}{\prod_j n_{\cdot j}!} \tag{8.1}$$

$$H_B(A)=\sum_j\frac{n_{\cdot j}}{N}\left\{\frac{1}{n_{\cdot j}}\log\frac{n_{\cdot j}!}{\prod_i n_{ij}!}\right\}$$

and

$$H_A(B)=\sum_i\frac{n_{i\cdot}}{N}\left\{\frac{1}{n_{i\cdot}}\log\frac{n_{i\cdot}!}{\prod_j n_{ij}!}\right\}$$

It is easily seen that

$$H(AB)=H(A)+H_A(B)=H(B)+H_B(A). \tag{8.2}$$

Note the formal identity of these equations and (1.8) (Section 1.7). We are now measuring (with Brillouin's index) the diversity of the occurrence matrix hierarchically, and doing it in two complementary ways.

In words, the meanings of the terms in (8.2) are as follows.

- $H(A)$ is the "species diversity" of the observed collection of N occurrences; (recall that the things counted are occurrences, not individual organisms).
- $H(B)$ is the habitat diversity of the collection of occurrences.
- $H_B(A)$ is the mean species diversity of occurrences within one kind of habitat averaged over all c kinds of habitats.
- $H_A(B)$ is the mean habitat diversity of occurrences of one species averaged over all r species.

One can regard $H_A(B)$ as a measure of the observed "mean niche width" of the r species; and $H_B(A)$ as a measure of their observed "mean niche overlap." Perhaps, as descriptive terms, "mean habitat span" and "mean habitat overlap" would be more precise. They are not independent, of course, nor should they be. To revert to niche terminology, if a given number of niches of given mean width are to be "packed" into a given space, then mean niche overlap is clearly subject to constraints. This assertion is intuitively obvious, though to derive the probability density of mean overlap under chosen hypotheses (for example, mutual independence of the niches) would be difficult.

To arrive at *standardized* measures of mean niche width and mean niche overlap, say W and L, which are independent of the relative abundances with which the different kinds of habitats, and the different species, chance to have been observed (equivalently, independent of the occurrence matrix's marginal totals), we may write

$$W=\frac{H_A(B)}{H(B)}$$

and (8.3)

$$L=\frac{H_B(A)}{H(A)}$$

with

$$0 \leq W < 1 \qquad \text{and} \qquad 0 \leq L < 1.$$

From (8.1) it will be found that if $n_{ij}=n_{i\cdot}n_{\cdot j}/N$ then (provided all n_{ij} are large) $W \simeq 1$ and $L \simeq 1$. Observe that we should have $\mathscr{E}(n_{ij})=n_{i\cdot}n_{\cdot j}/N$ if the two ways of classifying the occurrences were independent. If this were so the rows of the occurrence matrix would all be homogeneous with one another and with the row of column totals implying that the species occurred in the same relative proportions in all habitats; and the columns of the matrix would be homogeneous with one another and with the column of row totals implying that each kind of habitat contained all the species in the same relative proportions.

If each kind of habitat contains only one species, then $L=0$; when this is so there may be one habitat per species (so that $c=r$), or else some or all species may occur in more than one habitat (so that $c>r$). Likewise, if each species is found in only one habitat, then $W=0$; this may happen when there is a one-to-one correspondence between habitats and species so that $r=c$ and $L=0$; or else when some or all habitats contain more than one species so that $r>c$ and $L>0$. The fact that $\min(W)=0$ means that W might better be regarded as a measure of the amount by which mean niche width exceeds its minimum possible value rather than as a measure of mean niche width itself. To avoid prolixity, we shall represent the measures only by the symbols W and L as defined above and avoid giving names to them.

It is clear that, formally, we could take n_{ij}/N, $n_{i\cdot}/N$ and $n_{\cdot j}/N$ as estimates of probabilities and then use the Shannon index (cf Section 1.2) in place of the Brillouin index as a diversity measure. Then, using the symbolism introduced in Chapter 1 (H' for the Shannon index and H for the Brillouin index), (8.2) would remain true if primes were attached to all the the H's. Further, we could replace W and L, as defined in (8.3), by

$$W'=\frac{H'_A(B)}{H'(B)} \quad \text{and} \quad L'=\frac{H'_B(A)}{H'(A)}$$

with

$$0\leq W'\leq 1 \quad \text{and} \quad 0\leq L'\leq 1.$$

There are two strong objections to doing this, however. First, it entails the assumption that the N observations tabulated in an occurrence matrix are a truly random sample from some parent population of occurrences. In practice, this is hardly ever likely to be true and it is better, therefore, to treat the N occurrences as constituting a "collection" (in the sense of Section 1.4), and the occurrence matrix as a single multidimensional "observation." The second objection, which follows from the first, is that numerical values of L' and W' should, since they are assumed to be estimates of unknown "community" values, be accompanied by estimates of their standard errors (cf Section 1.4) and the standard errors have not yet been derived.

Therefore W and L as defined in (8.3) are preferable to W' and L' as standardized measures of mean niche width and overlap; W and L summarize properties of the particular observations at hand, and are thus automatically free of sampling error.

The problem of comparing mean niche widths and overlaps in different communities remains. It is not easy. It would be as meaningless to ask whether calculated values of W, for instance, from two different geographical regions differed "significantly" as it would be to ask whether, say, the heights of Peter and Paul differed "significantly"; either they differ or they do not, and unless we are trying to infer something about two unknown source populations (which, given a single observation from each, we cannot), problems of significance do not arise. Two or more occurrence matrices could be compared, provided r and c were the same for all, by the methods given in Sections 5.2 and 5.3 (and see Kullback, Kupperman and Ku, 1962) but only exceptionally will several occurrence matrices have the same r and c values.

Let us recall the purpose of calculating values of W and L. The values for a single occurrence matrix are uninteresting by themselves and usually the objective will be to search for trends, with geographic location, in the mean niche widths (relative to a group of related habitats) of a group of related species. To make progress in such an undertaking, one must obviously gather data from several geographic regions; the observations from each region will then give one occurrence matrix and one value of W and of L. Only when this has been done for several regions are trends likely to manifest themselves.

Examples of the calculation of W and L will be found in Pielou (1972, 1974*c*); the related species concerned were congeneric aphid species and their habitats congeneric plant species.

8.4 Diversity on Small Islands

Diversity, as measured by the number of species present, tends to be lower on oceanic islands than in equal areas of continental mainland with similar environments. This fact led MacArthur and Wilson (1967) to their now well known and generally accepted theory of island biogeography. As they pointed out, any island receives new immigrant species from the nearest mainland from time to time; and loses species, by local extinction, from time to time. When the rates of immigration and extinction are in balance, then the number of species in the island will be in dynamic equilibrium at a roughly constant level determined by the relative magnitudes of these rates. Among predictions suggested by the theory are that, given islands of equal size, the number of species on an island should vary inversely with its distance from the mainland that serves as faunal and floral source for the island; this follows since the rate of immigration of new species must fall off with increasing distance from the source whereas the extinction rate is independent of it. And, given islands equidistant from the source, the number of species should vary directly with island size, since the larger the island, the lower its local extinction rate and the higher its immigration rate. MacArthur and Wilson (1967) collected much evidence to support their theory; in particular, they used data on the numbers of breeding birds in various islands and archipelagoes of the western Pacific.

Detailed experimental studies on the colonization of islands have been made by Wilson and Simberloff; (see Wilson and Simberloff, 1969; Simberloff and Wilson, 1969; Simberloff, 1969; and Wilson, 1969). Their experimental islands were clumps of red mangrove trees growing in shallow water in Florida Bay. The diameters of the islands ranged from 10 m to 25 m, and they were at distances from the mainland (the southern tip of Florida) ranging from 2 m to 1200 m. The terrestrial animals living as permanent inhabitants of these islands are all arboreal arthropods (insects, spiders, scorpions, pseudoscorpions, centipedes, millipedes and sowbugs) and the experimental islands were completely "defaunated" by fumigation at the beginning of the experiment. The process of recolonization of each island was then monitored at frequent intervals for many months.

Simberloff's (1969) theoretical model for the recolonization process is as follows.*

Consider one given species and one empty island. For the species to colonize the island, at least one propagule of the species must reach the island; a propagule is the minimum number of individuals required to found a new population and is usually (for an animal species) a gravid female or else a pair of individuals of opposite sex.

Suppose that, during a short time interval of length Δt, the probability that a propagule of the species under consideration will reach the island is $\iota\,\Delta t$; (this probability is unaffected by whether the species is already present in the island or not). Suppose also, that if the species is already a member of the island's fauna, the probability that it will die out (become locally extinct in the island) during Δt is $\varepsilon\,\Delta t$. Assume the interval Δt to be so short that the probability is vanishingly small that both these events will occur in a single interval. Let $p(t)$ be the probability that the species is present in the island at time t. Then

$$p(t+\Delta t)=p(t)(1-\varepsilon\,\Delta t)+[1-p(t)]\iota\,\Delta t. \tag{8.4}$$

Now define an indicator variable $s(t)$ by putting

$$s(t)=\begin{cases}1 & \text{if the species is in the island at time } t\\ 0 & \text{if the species is not in the island at time } t.\end{cases}$$

Then the expectation of $s(t)$ is

$$\mathscr{E}[s(t)]=\Pr[s(t)=1]. \tag{8.5}$$

Substituting from (8.5) into (8.4) and rearranging gives

$$\mathscr{E}[s(t+\Delta t)]=\iota\,\Delta t+\mathscr{E}[s(t)](1-\varepsilon\,\Delta t-\iota\,\Delta t).$$

Therefore

$$\lim_{t\to\infty}\frac{\mathscr{E}[s(t+\Delta t)]-\mathscr{E}[s(t)]}{\Delta t}=\frac{d\,\mathscr{E}[s(t)]}{dt}=\iota-(\varepsilon+\iota)\,\mathscr{E}[s(t)].$$

* In the condensed account of part of Simberloff's (1969) model given here it has been possible to economize on symbols. Therefore those used here are not identical with those in his paper.

Integrating, we obtain

$$\iota-(\varepsilon+\iota)\,\mathscr{E}[s(t)]=K\exp[-(\varepsilon+\iota)t],$$

where K is a constant of integration. It may be evaluated by noting that at $t=0$ the species was absent from the island by hypothesis. Thus

$$\mathscr{E}[s(0)]=0$$

and hence $K=\iota$. Therefore

$$\mathscr{E}[s(t)]=\frac{\iota}{\varepsilon+\iota}\{1-\exp[-(\varepsilon+\iota)t]\}$$

and

$$\lim_{t\to\infty}\mathscr{E}[s(t)]=\frac{\iota}{\varepsilon+\iota}$$

Now recall that $\mathscr{E}[s(t)]$ is the expectation of an indicator variable that takes the value 1 or 0 according as the given species is or is not present in the island. If there is a pool of P species in the "source" from which immigrants come, if the respective values of ι and ε are the same for all these species, and if the species are all independent of one another, then $S(t)$, the number of species in the island at time t, has expectation

$$\mathscr{E}[S(t)]=\mathscr{E}[Ps(t)]=\frac{P\iota}{\varepsilon+\iota}\{1-\exp[-(\varepsilon+\iota)t]\} \qquad (8.6)$$

and

$$\lim_{t\to\infty}\mathscr{E}[S(t)]=\frac{P\iota}{\varepsilon+\iota}=\check{S}, \qquad \text{say.} \qquad (8.7)$$

Thus

$$\mathscr{E}[S(t)]=\check{S}\{1-\exp[-(\varepsilon+\iota)t]\}. \qquad (8.8)$$

Consider also the time required for the number of species (still assumed all to have the same values of ι and ε) colonizing an initially empty island to reach a fraction x of its ultimate equilibrium value $\check{S}$. Denote this time interval by t_x so that

$$\mathscr{E}[S(t_x)]=\check{S}x.$$

Then, from (8.8),

$$x=1-\exp[-(\varepsilon+\iota)t_x]$$

whence

$$t_x=\frac{-\ln(1-x)}{(\varepsilon+\iota)}. \qquad (8.9)$$

Although in the foregoing discussion it has been assumed that ι and ε are the same for all species, the model's predictions will be qualitatively the same even when this restriction is relaxed. Among these predictions are: (1) that the number of species in an initially empty offshore island will increase at a steadily decreasing rate, approaching the equilibrium level $\check{S}$ asymptotically; and (2) that, since values of ι will tend to be smaller the farther away an island lies from the source continent, a far island will tend to have a lower $\check{S}$ value than a near island of the same size and will tend to gain the equilibrium quota of species at a slower rate; (see the solid curves in Figure 8.2).

Now recall that the foregoing discussion presupposes that the species colonizing an island do not interact with one another. That is, they are assumed to be independent: competitive exclusion does not take place. For this reason, Wilson and Simberloff (1969) have called $\check{S}$ the *noninteractive equilibrium* number of species in an island; they expect the actual number—the *interactive equilibrium* number, say $\check{S}_1$—to be somewhat lower. Their experimental results support this prediction. It was found repeatedly that, on an experimentally defaunated island, the species number first rose to a value greater than the predefaunation level and then

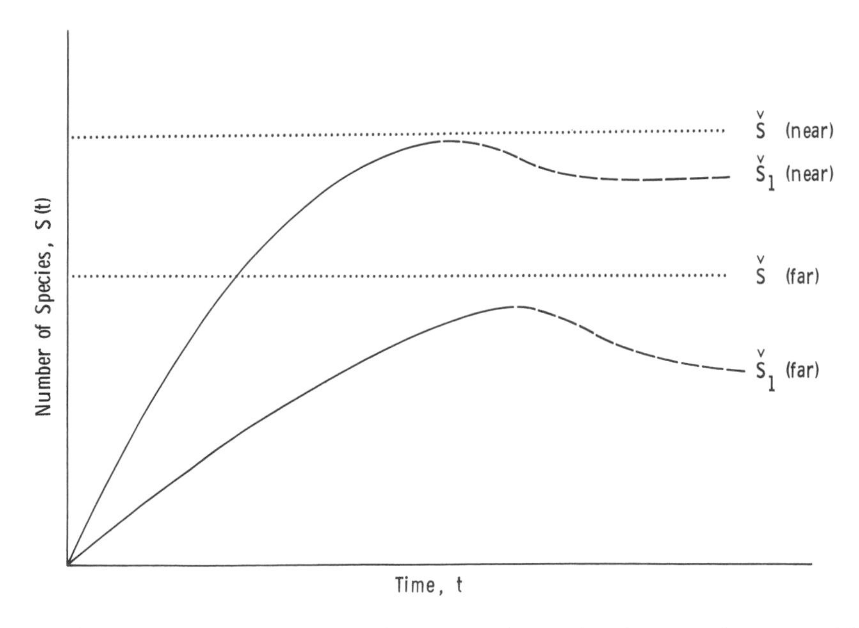

Figure 8.2 Representative curves showing how species numbers increase when near and far defaunated islands are recolonized. Solid lines: $S(t)$ versus t while populations are small and not interacting. Broken lines: $S(t)$ versus t after populations have begun to interact. (Redrawn from Simberloff, 1969.)

dropped back somewhat to a level that was fairly steadily maintained; (see Figure 8.2). Presumably the number rose to a value close to the non-interactive equilibrium level, $\check{S}$, to begin with, while species-populations were still small; and then, as the immigrant populations grew larger and began to interfere with one another, some of the competitively inferior species were excluded so that the number of species decreased to the interactive equilibrium level, $\check{S}_1$.

Notice that when $\check{S}_1$ species are present in interactive equilibrium, the equilibrium is assumed to be *dynamic*; that is, it is not assumed that an island will contain exactly the same complement of individual species indefinitely but merely that extinctions (resulting both from chance and from competitive exclusion) will be balanced by immigrations. This is equivalent to assuming that "blurring" is going on (see Section 6.2). For some species local extinction will be deferred because of the repeated chance arrivals of propagules of these species from the source region, and even when a weak competitor has been excluded its exclusion will only be temporary.

Now visualize what would happen if an island in dynamic interactive equilibrium (hence with $\check{S}_1$ species) were suddenly to be enclosed by a barrier that prevented the arrival of any propagules from outside. The number of species would presumably fall to a new steady value, say $\check{S}_2$, that may be called the *static interactive equilibrium* level. It is assumed that when this level has been reached all competitive exclusions have been completed and that therefore all the remaining coexisting species represent a *blended* rather than a *blurred* assemblage (cf Section 7.2). Immigration and extinction rates are again balanced since both are zero. Thus we assume that for only some of the $\check{S}_1$ species present at dynamic equilibrium is $\varepsilon > 0$; these are the species that are lost to an island when it is "fenced", either because of competitive inferiority or from chance. The remaining $\check{S}_2$ species have ε values so close to zero that the probability of their loss from the island in the foreseeable future is negligible. (Strictly speaking, of course, the static equilibrium is only a quasi-equilibrium since for any species extinction is always theoretically possible).

Nothing is at present known about the magnitude of the difference between $\check{S}_1$ and $\check{S}_2$ in different circumstances and for different taxocenes. For islands such as those examined by Wilson and Simberloff the difference is probably great; for comparatively large, oceanic islands it is no doubt quite small though whether it is ever virtually zero is unknown. The implications of this discussion from the point of view of conservation, and especially for the protection of rare or endangered species, are obvious. When land is cleared and forests destroyed to make room for the growing human population, the "islands" left as parks or reserves for the preservation of native floras and faunas must not be too small, nor too far from one another, otherwise they will eventually be found to contain only the $\check{S}_2$ "indestructible" species that are competitively tough enough, and prolific enough, to survive in static interactive equilibrium. Many of the species it was intended to conserve may be lost.

8.5 Temporal Changes in Diversity Through Geological Time

The fact that an observed community has a certain number of species does not imply, of course, that the number is not changing with time. There is a continuum of possible rates at which changes can occur, and changes at each of a number of different rates with as many different causes are usually all in progress simultaneously.

Three qualitatively different types of temporal change should be recognized, however. They are: (1) successional change such as occurs when a pioneer plant community on newly denuded land goes through progressive changes in composition until the stable climax is reached; (2) biotic changes caused by gradual changes in the world's climates and in the spatial arrangements of continents and oceans; and (3) the continous change caused by organic evolution.

With regard to (1) there is not much to be said beyond the fact that, so far as plant communities are concerned, species numbers increase (as, indeed, they must if they start from zero) in the early stages of a succession and then tend to decrease slightly from a subclimactic maximum as climax is finally reached.

With regard to (2) and (3), with which this section will be concerned, there is a great deal that could be said and here it will be possible only to give a very condensed account of some of the growing points of contemporary research.

Climatic change never ends but its effect on vegetation in the recent past (say the last ten thousand years) have been especially well studied. The total absence of vegetation from areas that were covered by ice at one time or another during the Pleistocene means that the latitudinal diversity gradient must necessarily have been much more pronounced then than it is at present; the possibility has been considered that the relative paucity of plant and animal species in high latitudes even now may be because areas that were under the Pleistocene ice sheets have still not fully recovered. This may well be a contributory cause, but is certainly not the only cause, of the gradient.

Evidence showing that a latitudinal gradient would exist even if there had been no glaciation is of two kinds. First is the fact that diversity gradients are not confined to regions that were ice-covered; they are equally strong in warm-temperate and tropical regions which were ice-free throughout the Quaternary; Stehli (1969) gives numerous examples and two of the clearest are shown here in Figure 8.3. Second is the fact that, as fossil evidence shows, diversity gradients existed long before the last ice age; for example (Stehli, Douglas and Newell, 1969), there appears to have been a clear latitudinal gradient (in the North Atlantic) in numbers of species of planktonic foraminifera during the Cretaceous (*circa* 75 million years b.p.) when the contrast in climate between low and high latitudes was much weaker than it is now and conditions everywhere were more equable. Thus a comparatively slight gradient in mean temperatures is enough to produce a latitudinal diversity gradient even without glaciation.

This is not to say, however, that the present strong diversity gradient is not to some extent augmented by the effects of past ice sheets. In fact we do not know to what extent changes in the steepness of the diversity gradient lag in their response to changes in the world's climatic gradients.

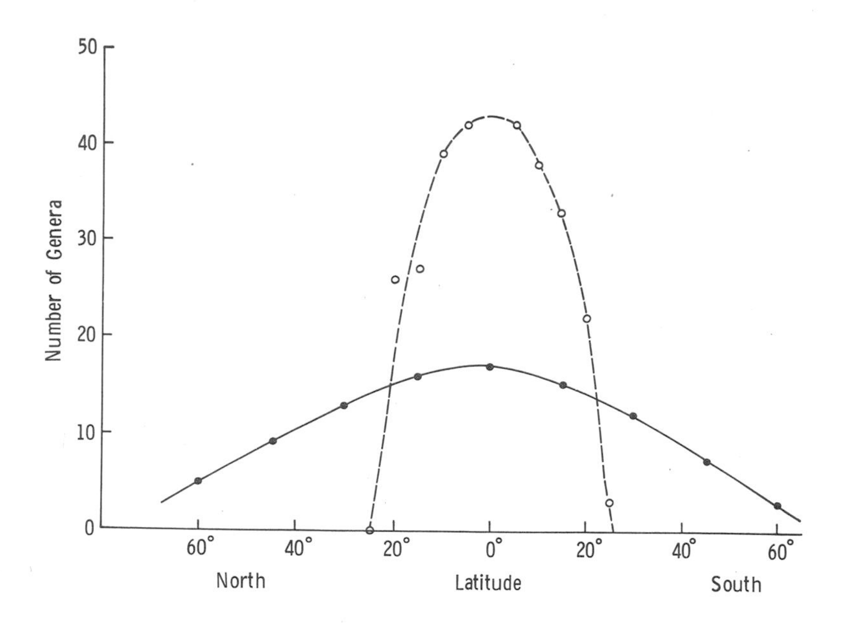

Figure 8.3 The variation of numbers of genera with latitude for mosquitoes (solid line) and butterflies of the family Danaidae (broken line). (Redrawn from Stehli, 1968.)

We turn now to larger areas and longer time periods. The way in which the animal kingdom throughout the world has diversified itself, by evolution, during the past 3×10^9 years has been reviewed by Simpson (1969). He considers the numbers of taxonomic orders known to have been extant in different geological periods, and gives the following outline of the way in which organic diversity has developed.

An enormous amount of diversification must have occurred between the dawn of life and the date of the beginning of the fossil record, and by the Ordovician period it appears that evolution of orders had proceeded to a point where aquatic (marine and freshwater) environments were ecologically saturated in the sense that what might be called "order packing" could go no farther. This hypothesis is supported by the fact that the number of aquatic orders remained roughly constant until the wave of faunal extinctions that occurred in Permian-Triassic time, which are attributed to severe climatic deterioration; thereafter (with an exception to be noted below) the number returned to the earlier level for the whole of the Cretaceous and Tertiary periods; although individual orders became extinct and were replaced by new ones, the numbers of orders in the different phyla did not vary greatly. The exception was the appearance, after the Permian-Triassic "low" in aquatic faunal diversity, of the teleost fishes, which added many new orders to the world's aquatic fauna.

Similarly, but much later, terrestrial communities developed; from the Silurian onwards, the number of terrestrial orders (chiefly insects and other nonaquatic arthropods, and amphibian and fully terrestrial vertebrates) increased; they reached a plateau and, presumably, ecological saturation, in the Permian. However, in the Tertiary, ordinal diversity in the terrestrial fauna increased markedly because of a rapid increase in the numbers of homoiothermic vertebrates (mammals and birds); these groups have a place in the history of terrestrial communities similar to that of the teleost fishes in aquatic communities. There was a concomitant decrease in the numbers of orders of reptiles and amphibians but the diversification of the warm-blooded vertebrates more than made up for it; moreover the flying vertebrates (of which only birds and a single order of bats now remain) simultaneously introduced an entirely new mode of life.

Thus worldwide faunal diversification has increased since life first appeared in a somewhat stepwise fashion, through the development and exploitation of adaptations permitting a succession of new modes of life. There have been setbacks, notably in the Permian-Triassic and again in the wave of mammal extinctions in the middle (pre-Pleistocene) Quaternary. Further, even though the numbers of orders appear to stabilize given enough time, this is not necessarily true of numbers of families. For example, the numbers of families in five different orders of reef-building corals have fluctuated enormously, with no sign of stabilization, from the Cambrian to the Quaternary. In contrast, there is evidence to believe that the numbers of families of land mammals in North and South America is currently in a state of dynamic stability.

The overall floral and faunal diversity of the whole world depends, of course, on the number of taxa (of whatever level is being considered) required to "saturate" the world's various environments, the number of different environments and, according to Valentine and Moores (1972), the "provinciality" of the world. This term connotes the degree to which the habitable world is fragmented so that each distinct type of environment is represented by many separate, small, mutually isolated parts each with an endemic set of plants and animals; that is, provinciality is defined as high when distinct floral and faunal provinces are numerous.

The way in which the continents drift, separate and reunite, so that the total land surface of the world has in some periods formed a single continent (Pangaea) and in other periods (as now) as many as six continents, has brought about continual changes in the degree of provinciality of the world's flora and fauna. Valentine and Moores (1972) discuss their theory in the context of marine shelf faunas. For marine organisms on continental shelves, separate faunal provinces will come into existence, and speciation will proceed independently in each, when the world's continental shelves are much subdivided and have barriers between the parts. For shelf organisms, continental land masses and deep ocean basins constitute such barriers to dispersal between provinces. In addition, provinciality will be enhanced if each continuous piece of shelf is further subdivided into climatic provinces; this effect will be most pronounced when (as now) most of the world's continental shorelines trend north and south, and when (again as now) the temperature gradients from poles to equator are steep.

Thus three factors promote provinciality: fragmentation of the continents; a prevalence of north-south shorelines; and strong climatic gradients. These factors, though not independent, cannot be counted upon to reinforce one another; however, at the present time it appears that they do, with the result that the present diversity of the world's plants and animals is (or was just before our species appeared) probably greater than it has ever been before. The very low marine diversity that characterized the Permian-Triassic boundary can be attributed to the low provinciality of the time; there was then a continuous continental shelf, devoid of any barriers to dispersal, surrounding the single continent, Pangaea. The same conditions—a single continent and low diversity—prevailed much earlier, in the Pre-cambrian and early Cambrian. Between these two diversity "lows," that is, during the Ordovician, Silurian, Devonian and Carboniferous periods, there were four continents and high diversity. From the late Triassic until the present, the continents have been drifting apart and diversity has been increasing.

These tectonically caused changes in the world's biotic diversity are superimposed on the steadily increasing diversity, which may or may not now be complete, that evolution has brought about. At a lower level, within-province diversity has been under the control of regional factors, especially environmental stability and productivity (cf Section 8.2). Factors determining global diversity can thus be thought of as operating at three hierarchical levels: the evolutionary, the tectonic, and the intra-provincial.

The interplay of these nested causes and the responses of natural communities to the often contradictory commands of the inanimate world offer a persistent challenge to research workers. Diversity waxes and wanes at different levels—different levels in the hierarchy of spatial areas, different levels in the hierarchy of time periods, and different levels in the taxonomic hierarchy. All investigators are familiar with the way in which a theory that fully and satisfyingly explains one phenomenon may prove on reflection to be a refutation of an earlier, equally good theory that explained another phenomenon. This, itself a well-known phenomenon in the behaviour of theories, is unlikely to stop, and we can only keep on working.

Bibliography

Abbott, R. T. (1968). *Seashells of North America.* Golden Press, New York.

Abele, L. G. (1974). Species diversity of decapod crustaceans in marine habitats. *Ecology* **55**: 156–161.

Aitchison, J. and **J. A. C. Brown** (1966). *The Lognormal Distribution.* Cambridge University Press.

Anscombe, F. J. (1950). Sampling theory of the negative binomial and logarithmic series distributions. *Biometrika* **37**: 358–382.

Ashmole, N. P. (1968). Body size, prey size and ecological segregation in five sympatric tropical terns (Aves: Laridae). *Syst. Zool.* **17**: 292–304.

Barton, D. E. and **F. N. David** (1959). The dispersion of a number of species. *J. Roy. Statist. Soc., **B,*** **21**: 190–194.

Basharin, G. P. (1959). On a statistical estimate for the entropy of a sequence of independent random variables. *Theory of Probability and Its Applications* **4**: 333–336.

Birch, M. W. (1963). An algorithm for the logarithmic series distribution. *Biometrics* **19**: 651–652.

Bliss, C. I. (1965). An analysis of some insect trap records. In "Classical and Contagious Discrete Distributions" (G. P. Patil, Ed.). Statistical Publishing Society, Calcutta.

Brian, M. V. (1953). Species frequencies in random samples from animal populations. *J. Anim. Ecol.* **22**: 57–64.

Briggs, D. and **S. M. Walters** (1969). *Plant Variation and Evolution.* McGraw-Hill, New York.

Brillouin, L. (1962). *Science and Information Theory.* 2nd ed. Academic Press, New York.

Bulmer, M. G. (1974). On fitting the Poisson lognormal distribution to species-abundance data. *Biometrics* **30**: 101–110.

Cain, A. J. (1969). Speciation in tropical environments: summing up. In "Speciation in Tropical Environments" (R. H. Lowe-McConnell, Ed.). Academic Press, New York.

Chapman, A. R. O. (1973). A critique of prevailing attitudes towards the control of seaweed zonation on the seashore. *Botanica Marina* **16**: 80–82.

Clarke, B. (1969). The evidence for apostatic selection. *Heredity* **24**: 347–352.

Clayton, W. D. (1966). Vegetation ripples near Gummi, Nigeria. *J. Ecol.* **54**: 415–417.

Cohen, A. C. Jr. (1961). Tables for maximum likelihood estimates: singly truncated and singly censored samples. *Technometrics* **3**: 535–541.

Cohen, J. E. (1968). Alternate derivations of a species-abundance relation *Amer. Natur.* **102**: 165–172.

Connell, J. H. (1961). The influence of interspecific competition and other factors on the distribution of the barnacle *Chthamalus stellatus. Ecology* **42**: 710–723.

Connell, J. H. and **E. Orias** (1964). The ecological regulation of species diversity. *Amer. Natur.* **98**: 399–414.

Conover, W. J. (1971). *Practical Nonparametric Statistics.* Wiley, New York.

Cox, D. R. (1962). *Renewal Theory.* Wiley, New York.

Culver, D. C. (1970). Analysis of simple cave communities. I. Caves as islands. *Evolution* **24**: 463–474.

DeBenedictis, P. A. (1973). On the correlations between certain diversity indices. *Amer. Natur.* **107**: 295–302.

Diamond, J. M. (1973). Distributional ecology of New Guinea Birds. *Science* **179**: 759–769.

Eberhardt. L. L. (1969). Some aspects of species diversity models. *Ecology* **50**: 503–505.

Ehrlich, P. R. and **P. H. Raven** (1969). Differentiation of populations. *Science* **165**: 1228–1232.

Eltringham, S. K. (1971). *Life in Mud and Sand.* English Universities Press.

Feller, W. (1966). *An Introduction to Probability Theory and Its Applications.* Vol. II. Wiley, New York.

Fisher, R. A., A. S. Corbet and **C. B. Williams** (1943). The relation between the number of species and the number of individuals in a random sample of an animal population. *J. Anim. Ecol.* **12**: 42–58.

Goldman, S, (1953). Some fundamentals of information theory. In "Information Theory in Biology" (H. Quastler, Ed.). University of Illinois Press, Urbana.

Harper, J. L. (1967). A Darwinian approach to plant ecology. *J. Ecol.* **55**: 247–270.

Harper, J. L. (1969). The role of predation in vegetational diversity. In "Diversity and Stability in Ecological Systems" (G. M. Woodwell and H. H. Smith, Eds.). Brookhaven Symposium in Biology, No. 22.

Harper, J. L., J. N. Clatworthy, I. H. McNaughton and **G. R. Sagar** (1961). The evolution and ecology of closely related species living in the same area. *Evolution* **15**: 209–227.

Heyer, W. R. and **K. A. Berven** (1973). Species diversities of herpetofaunal samples from similar microhabitats at two tropical sites. *Ecology* **54**: 642–645.

Holgate, P. (1969). Species frequency distributions. *Biometrika* **56**: 651–660.

Holmes, R. T. and **F. A. Pitelka** (1968). Food overlap among coexisting sandpipers on northern Alaskan tundra. *Syst. Zool.* **17**: 305–318.

Horn, H. S. and **R. H. MacArthur** (1972). Competition among fugitive species in a harlequin environment. *Ecology* **53**: 749–752.

Hosie, R. C. (1969). *Native Trees of Canada.* Canadian Forestry Service.

Hurlbert, S. H. (1971). The nonconcept of species diversity: a critique and alternative parameters. *Ecology* **52**: 577–586.

Hutchinson, G. E. (1957). Concluding remarks. *Cold Spring Harbor Symp. Quant. Biol.* **22**: 415–427.

Hutchinson, G. E. (1965). *The Ecological Theater and the Evolutionary Play.* Yale University Press, New Haven.

Hutchinson, G. E. (1967). *A Treatise on Limnology.* Vol. II. Wiley, New York.

Janzen, D. H. (1967). Why mountain passes are higher in the tropics. *Amer. Natur.* **101**: 233–249.

Janzen, D. H. (1970). Herbivores and the number of tree species in tropical forests. *Amer. Natur.* **104**: 501–528.

Kendall, M. G. (1948). *Rank Correlation Methods.* Griffin, London.

Khamis, S. H. and **W. Rudert** (1965). *Tables of the Incomplete Gamma Function Ratio.* Justus von Liebig Verlag, Darmstadt.

Khinchin, A. I. (1957). *Mathematical Foundations of Information Theory.* Dover, New York.

King, C. E. (1964). Relative abundance of species and MacArthur's model. *Ecology* **45**: 716–727.

Kullback, S., M. Kupperman, and **H. H. Ku** (1962). Tests for contingency tables and Markov chains. *Technometrics* **4**: 573–608.

Leslie, P. H. (1958). A stochastic model for studying the properties of certain biological systems by numerical methods. *Biometrika* **45**: 16–31.

Levins, R. (1968). Toward an evolutionary theory of the niche. In "Evolution and Environment" (E. T. Drake Ed.). Yale University Press, New Haven.

Levins, R. and **D. Culver** (1971). Regional coexistence of species and competition between rare species. *Proc. Nat. Acad. Sci. USA*. **68**: 1246–1248.

Lloyd, M., R. F. Inger and **F. W. King** (1968). On the diversity of reptile and amphibian species in a Bornean rain forest. *Amer. Natur.* **102**: 497–515.

MacArthur, R. H. (1957). On the relative abundance of bird species. *Proc. Nat. Acad. Sci. Wash.* **43**: 293–295.

MacArthur, R. H. (1958). Population ecology of some warblers of north-eastern coniferous forests. *Ecology* **39**: 599–619.

MacArthur, R. H. and **R. Levins** (1967). The limiting similarity, convergence and divergence of coexisting species. *Amer. Natur.* **101**: 377–385.

MacArthur, R. H. and **E. O. Wilson** (1967). *The Theory of Island Biogeography*. Princeton University Press, Princeton.

Margalef, D. R. (1958). Information theory in ecology. *Gen. Syst.* **3**: 36–71.

May, R. M. (1973). *Stability and Complexity in Model Ecosystems.* Princeton University Press, Princeton.

May, R. M. (1975). Patterns of species abundance and diversity. (in press)

Maynard Smith, J. (1974). *Models in Ecology*. Cambridge University Press, London.

Mayr. E. (1963). *Animal Species and Evolution.* Belknap Press, Harvard University, Boston.

Miller, R. S. (1969). Competition and species diversity. In "Diversity and Stability in Ecological Systems (G. M. Woodwell and H. H. Smith Eds.). Brookhaven Symposium in Biology No. 22.

Nelson, W. C. and **H. A. David** (1967). The logarithmic distribution: a review. *Virginia Journal of Sci.* **18**: 95–102.

O'Neill, R. V. (1967). Niche segregation in seven species of diplopods. *Ecology* **48**: 983.

Paine, R. T. (1966). Food web complexity and species diversity. *Amer. Natur.* **100**: 65–75.

Patrick, R. (1968). The structure of diatom communities in similar ecological conditions. *Amer. Natur.* **102**: 173–183.

Paulson, D. R. (1973). Predator polymorphism and apostatic selection. *Evolution* **27**: 269–277.

Pearson, E. S. and **H. O. Hartley** (1954). *Biometrika Tables for Statisticians.* Vol. I. Cambridge University Press.

Pianka, E. R. (1966). Latitudinal gradients in species diversity: a review of concepts. *Amer. Natur.* **100**: 33–46.

Pianka, E. R. (1970). On *r*- and *K*-selection. *Amer. Natur.* **104**: 592–597.

Pielou, D. P. and **A. N. Verma** (1968). The arthropod fauna associated with the birch bracket fungus, *Popyporus betulinus*, in eastern Canada. *Can. Entomol.* **100**: 1179–1199.

Pielou, E. C. (1966*a*). Species-diversity and pattern-diversity in the study of ecological succession. *J. Theoret. Biol.* **10**: 370–383.

Pielou, E. C. (1966*b*). The measurement of diversity in different types of biological collection. *J. Theoret. Biol.* **13**: 131–144.

Pielou, E. C. (1967). A test for random mingling in the phases of a mosaic. *Biometrics* **23**: 657–670.

Pielou, E. C. (1969). *An Introduction to Mathematical Ecology*. Wiley, New York.

Pielou, E. C. (1971). Measurement of structure in animal communities. In "Ecosystem Structure and Function" (J. A. Wiens, Ed.). *Proc. 31st Annual Biology Colloquim*. Oregon State U.P., Portland.

Pielou, E. C. (1972). Niche width and overlap: a method for measuring them. *Ecology* **53**: 687–692.

Pielou, E. C. (1974*a*). Competition on an environmental gradient. In "Mathematical Problems in Biology" (P. van den Driessche, Ed.). Springer-Verlag.

Pielou, E. C. (1974*b*). Vegetation zones: repetition of zones on a monotonic environmental gradient. *J. Theoret. Biol.* 47: 485–489.

Pielou, E. C. (1974*c*). Biogeographic range comparisons and evidence of geographic variation in host parasite relations. *Ecology* 55: 1359–1367.

Pielou, E. C. (1974d). *Population and Community Ecology: Principles and Methods*. Gordon and Breach, New York.

Pielou, E. C. (1975). Ecological models on an environmental gradient. *Proc. Symp. Applic. of Statistics*. (R. P. Gupta, Ed.). North-Holland American Elsevier.

Pielou, E. C. and **A. N. Arnason** (1965). Correction to one of MacArthur's species-abundance formulas. *Science* **151**: 592.

Pimentel, D., E. H. Feinberg, P. W. Wood and **J. T. Hayes** (1965). Selection, spatial distribution, and the coexistence of competing fly species. *Amer. Natur.* **99**: 97–109.

Poulson T. L. and **D. C. Culver** (1969). Diversity in terrestrial cave communities. *Ecology* **50**: 153–158.

Preston, F. W. (1948). The commonness and rarity of species. *Ecology* **29**: 254–283.

Preston, F. W. (1962). The canonical distribution of commonness and rarity, Parts I and II. *Ecology* **43**: 185–215 and 410–432.

Renyi, A. (1961). On measures of entropy and information. *Proc. 4th. Berkeley Symposium on Math. Stat. and Prob.* **1**: 547–561.

Richards, P. W. (1969). Speciation in the tropical rain forest and the concept of the niche. In "Speciation in Tropical Environments" (R. H. Lowe McConnell, Ed.). Academic Press, New York.

Rothstein, S. I. (1973). The niche-variation model—is it valid? *Amer. Natur.* **107**: 598–620.

Sanders, H. L. (1968). Marine benthic diversity: a comparative study. *Amer. Natur.* **102**: 243–282.

Sanders, H. L. (1969). Benthic marine diversity and the stability-time hypothesis. In "Diversity and Stability in Ecological Systems" (G. M. Woodwell and H. H. Smith, Eds.). Brookhaven Symposium in Biology No. 22.

Shannon, C. E. and **W. Weaver** (1949). *The Mathematical Theory of Communication.* University of Illinois Press, Urbana.

Siegal, S. (1956). *Nonparametric Statistics for the Behavioral Sciences.* McGraw Hill, New York.

Simberloff, D. S. (1969). Experimental zoogeography of islands. A model for insular colonization. *Ecology* **50**: 296–314.

Simberloff, D. S. and **E. O. Wilson** (1969). Experimental zoogeography of islands. The colonization of empty islands. *Ecology* **50**: 278–296.

Simpson, E. H. (1949). Measurement of diversity. *Nature* **163**: 688.

Simpson, G. G. (1964). Species density of North American Recent mammals. *Syst. Zool.* **13**: 57–73.

Simpson, G. G. (1969). The first three billion years of community evolution. In "Diversity and Stability in Ecological Systems" (G. M. Woodwell and H. H. Smith, Eds.). Brookhaven Symposium in Biology No. 22.

Skellam, J. G. (1951). Random dispersal in theoretical populations. *Biometrika* **38**: 196–218.

Slobodkin, L. B. and **L. Fishelson** (1974). The effect of the cleaner-fish, *Labroides dimidiatus* on the point diversity of fishes on the reef front at Eilat. *Amer. Natur.* **108**: 369–376.

Smith, F. E. (1972). Spatial heterogeneity, stability and diversity in ecosystems. In "Growth by Intussusception. Ecological Essays in Honor of Evelyn Hutchinson. (E. S. Deevey, Ed.) *Trans. Conn. Acad. Arts and Sci.* **44**.

Stehli, F. G. (1968). Taxonomic diversity gradients in pole location: the Recent model. In "Evolution and Environment" (E. T. Drake, Ed.). Yale University Press, New Haven.

Stehli, F. G., R. G. Douglas and **N. D. Newell** (1969). Generation and maintenance of gradients in taxonomic diversity. *Science* **164**: 947–950.

Strobeck, C. (1973). *N* species competition. *Ecology* **54**: 650–654.

Thorson, G. (1950). Reproduction and larval ecology of marine bottom invertebrates. *Biol. Rev.* **25**: 1–45.

Thorson, G. (1957). Bottom communities (sublittoral and shallow shelf). In "Treatise of Marine Ecology and Paleoecology" (H. S. Ladd, Ed.). *Geol. Soc. Amer. Mem.* **67**.

Valentine, J. W. and **E. M. Moores** (1972). Global tectonics and the fossil record. *J. Geol.* **80**: 167–184.

Wangersky, P. J. (1972). Evolution and the niche concept. In "Growth by

Intussusception. Ecological Essays in Honor of Evelyn Hutchinson" (E. S. Deevey, Ed.). *Trans. Conn. Acad. Arts and Sci.* **44**.

Wangersky, P. J. and **W. J. Cunningham** (1956). On time lags in equations of growth. *Proc. Nat. Acad. Sci. Wash.* **42**: 699–702.

Washburn, A. L. (1956). Classification of patterned ground and review of suggested origins. *Bull. Geol. Soc. Amer.* **67**: 823–866.

Watt, A. S., R. M. S. Perrin, and **R. G. West** (1966). Patterned ground in Breckland: structure and composition. *J. Ecol.* **54**: 239–258.

Weaver, W. (1948). Probability, rarity, interest, and surprise. *Sci. Monthly* **67**: 390–392.

West, R. G. (1968). *Pleistocene Geology and Biology*. Longmans, Green.

Whittaker, R. H. (1972). Evolution and measurement of species diversity. *Taxon* **21**: 213–251.

Whitworth, W. A. (1934). *Choice and Chance*. Steichert, New York.

Wiegert, R. G. (1974). Litterbag studies of microarthropod populations in three South Carolina old fields. *Ecology* **55**: 94–102.

Wilks, S. S. (1962). *Mathematical Statistics*. Wiley, New York.

Williams, C. B. (1964). *Patterns in the Balance of Nature*. Academic Press, N.Y.

Williams, L. G. (1964). Possible relationships between plankton-diatom species numbers and water quality estimates. *Ecology* **45**: 809–823.

Wilson, E. O. (1969). The species equilibrium. In "Diversity and Stability in Ecological Systems" (G. M. Woodwell and H. H. Smith, Eds.). Brookhaven Symposium in Biology No. 22.

Wilson, E. O. and **D. S. Simberloff** (1969). Experimental zoogeography of islands. Defaunation and monitoring techniques. *Ecology* **50**: 267–277.

Author Index

Subject Index